T0094291

Climate Change Discourse in Russia

This book explores the development of climate change discourses in Russia. It contributes to the study of climate change as a cultural idea by developing the extensive Anglophone literature on environmental science, politics and policy pertaining to climate change in the West to consider how Russian discourses of climate change have developed. Drawing on contributors specialising in numerous periods, regions, disciplines and topics of study, the central thread of this book is the shared attempt to understand how environmental issues, particularly climate change, have been understood, investigated and conceptualised in Soviet and post-Soviet Russia. The chapters aim to complement work on the history of the discursive political construction of climate change in the West by examining a highly contrasting (but intimately related) cultural context. Russia remains one of the world's largest greenhouse gas emitters with one of the most carbon-intensive economies. As the world begins to suffer the extreme consequences of anthropogenic climate change, finding adequate solutions to global environmental problems necessitates the participation of all countries. Russia is a central actor in this global process and it, therefore, becomes increasingly important to understand climate change discourse in this region. Insights gained in this area may also be illuminating for examining environmental discourses in other resource rich regions of the world with alternative economic and political experiences to that of the West (e.g. China, the Middle East).

This book will be of great interest to students and scholars of Russian environmental policy and politics, climate change discourses, environmental communication and environment and sustainability in general.

Marianna Poberezhskaya is Lecturer in International Relations and Politics at Nottingham Trent University, UK.

Teresa Ashe is Associate Lecturer and Honorary Fellow of the Geography Department at the Open University, UK.

Routledge Focus on Environment and Sustainability

The Application of Science in Environmental Impact Assessment
By Aaron Mackinnon, Peter Duinker and Tony Walker

Jainism and Environmental Philosophy
Karma and the Web of Life
By Aidan Rankin

Social Sustainability, Climate Resilience and Community-Based Urban Development
What About the People?
By Cathy Baldwin and Robin King

South Africa's Energy Transition
A Roadmap to a Decarbonised, Low-cost and Job-rich Future
By Terence Creamer and Tobias Bischof-Niemz

The Environmental Sustainable Development Goals in Bangladesh
Edited by Samiya A. Selim, Shantanu Kumar Saha, Rumana Sultana and Carolyn Roberts

Climate Change Discourse in Russia
Past and Present
Edited by Marianna Poberezhskaya and Teresa Ashe

The Greening of US Free Trade Agreements
From NAFTA to the Present Day
Linda J. Allen

For more information about this series, please visit: www.routledge.com/ Routledge-Focus-on-Environment-and-Sustainability/book-series/RFES

Climate Change Discourse in Russia

Past and Present

**Edited by
Marianna Poberezhskaya
and Teresa Ashe**

Routledge
Taylor & Francis Group

LONDON AND NEW YORK

earthscan
from Routledge

First published 2019
by Routledge
2 Park Square, Milton Park, Abingdon, Oxon OX14 4RN

and by Routledge
711 Third Avenue, New York, NY 10017

Routledge is an imprint of the Taylor & Francis Group, an informa business

British Library Cataloguing-in-Publication Data
A catalogue record for this book is available from the British Library

Library of Congress Cataloging-in-Publication Data
Names: Poberezhskaya, Marianna, editor. | Ashe, Teresa, editor.
Title: Climate change discourse in Russia : past and present / edited by Marianna Poberezhskaya and Teresa Ashe.
Description: Abingdon, Oxon ; New York, NY : Routledge, 2019. | Series: Routledge focus on environment and sustainability
Identifiers: LCCN 2018022761 | ISBN 9781138493209 (hardback) | ISBN 9781351028660 (ebook)
Subjects: LCSH: Climatic changes—Russia (Federation) | Climatic changes—Social aspects—Russia (Federation)—History.
Classification: LCC QC903.2.R8 C55 2019 | DDC 363.738/740947—dc23
LC record available at https://lccn.loc.gov/2018022761

ISBN: 978-1-138-49320-9 (hbk)
ISBN: 978-1-351-02866-0 (ebk)

Typeset in Times New Roman
by Apex CoVantage, LLC

Contents

Figures and Table

Figures

Table

Contributors

Teresa Ashe is Associate Lecturer and Honorary Fellow of the Geography Department at the Open University, UK. She teaches social science courses in politics, economics and the environment and researches the topics of climate change discourse, climate communication and climate scepticism. She is currently working on a book titled *Climate Change: Discourses of Making and Un-Making*, which looks at the development of climate change politics with a particular interest in the emergence of climate change scepticism in the US. Her research is particularly concerned with how technical scientific discourse, whether natural scientific or economic, holds power in environmental politics.

Benjamin Beuerle is a postdoctoral research fellow at the German Historical Institute Moscow, Russia. He has published a number of articles on legislative reform debates and politics in late imperial Russia and is the author of *Russlands Westen. Westorientierung und Reformgesetzgebung im ausgehenden Zarenreich, 1905–1917*, Harrassowitz (2016). He is the scientific coordinator of a new joint research project on Russia in the Asia-Pacific region developed at the German Historical Institute Moscow and an associate member of the research project 'Soviet Climate Science and its Intellectual Legacies'. His current research interests include Russian attitudes and political approaches to climate change, urban air pollution and renewable energy projects from late Soviet times until today.

Katja Doose is a research fellow at the department for Russian Environmental Studies, University of Birmingham, UK. She holds a PhD in Eastern European History from the University of Tübingen. From 2015–2017, Katja has been a fellow of the ZEIT-Foundation. Her research focuses on the environmental history of the Soviet Union, including disaster studies and climate change science as well as urban history. She recently published 'The Armenian Earthquake of 1988. The perfect Stage for the Karabakh Conflict', *Europe-Asia Studies* (2018); 'Eco-nationalism or

Environmental Legitimacy? The ecological transition of the Armenian Communist Party 1956–1991', *Ab Imperio* (2018); and 'Spitak: the last petrified Soviet Utopia', Tigran Harutyunyan, *Architectural Guide Yerevan* (2017).

Ellie Martus is a WIRL-Marie Skłodowska-Curie COFUND Fellow at the Institute of Advanced Study, University of Warwick, UK. Ellie's research explores the politics of the environment and climate in the post-Soviet region. She completed her PhD in politics at the University of NSW and an MPhil in Russian and East European Studies at St Antony's College, Oxford. Ellie has been a visiting fellow at the Australian National University's Centre for European Studies and is currently a visiting researcher at the International Center for the Study of Institutions and Development, Higher School of Economics, Moscow.

Anna Mazanik is a visiting professor at the History Department of Central European University and an archivist at the Blinken Open Society Archives (Hungary), where she curates the Soviet collections. Anna holds a PhD in comparative history from CEU. She has also studied at Moscow State University, Johns Hopkins University and the Rachel Carson Center for Environment and Society. Before joining the OSA in 2016, she was a postdoctoral fellow at the Herder Institute, New Europe College, and the Max Planck Institute for the History of Science. Her scholarly work focuses on environmental history, urban history and public health in imperial and Soviet Russia.

Jonathan Oldfield is a reader in Russian Environmental Studies at the University of Birmingham, UK. He is currently working on the Soviet understanding of climate science. Jonathan's research is funded by the ESRC, AHRC and British Academy. He recently published *The Development of Russian Environmental Thought: Scientific and Geographical Perspectives on the Natural Environment* (with D. Shaw), Routledge (2016); 'Imagining climates past, present and future: Soviet contributions to the science of anthropogenic climate change, 1953–1991', *Journal of Historical Geography* (2018); and 'Mikhail Budyko's (1920–2001) contributions to Global Climate Science: from heat balances to climate change and global ecology', *WIREs Climate Change*, (2016).

Marianna Poberezhskaya is a lecturer in International Relations, Nottingham Trent University, UK, with research interests in environmental communication, post-Soviet climate politics and the political economy of mass media. She received her PhD in Politics and IR from the University of Nottingham, UK (2013), which was followed by a Postdoctoral Lecturing Fellowship at the University of East Anglia, UK. Marianna's recent publications include 'Blogging about Climate Change in Russia: Activism, Scepticism and Conspiracies', *Environmental Communication*, (2017), *Communicating Climate Change in Russia: state and propaganda*, Routledge (2016) and 'Measuring and modeling Russian newspaper coverage of climate change' (with C. Boussalis and T. Coan), *Global Environmental Change*, 41 (2016).

Veli-Pekka Tynkkynen is an associate professor in Russian Environmental Studies at the Aleksanteri Institute and the Helsinki Institute of Sustainability Science, University of Helsinki, Finland. His research fields are energy and environmental policies, energy security, natural resources, political/societal power and culture in Russia. His recent publications include: Tynkkynen, V-P. et al. (eds.) *Russia's Far North: The Contested Energy Frontier*, Routledge (2018); 'Climate Denial revisited: (Re)contextualising Russian Public Discourse on Climate Change during Putin 2.0' (with N. Tynkkynen), *Europe-Asia Studies*, (forthcoming 2018); and 'Energy as Power – Gazprom, Gas Infrastructure, and Geo-Governmentality in Putin's Russia', *Slavic Review* 75, 2 (2016).

Abbreviations

AANII	Arctic and Antarctic Research Institute
APG	Associated Petroleum Gas
APEC	Asia-Pacific Economic Cooperation
AR	Annual Report
BAM	Baikal-Amur Mainline
BP	British Petroleum
BRICS	Brazil, Russia, India, China and South Africa
CDP	Carbon Disclosure Project
CIA	Central Intelligence Agency
CO_2	Carbon dioxide
CSR	Corporate Social Responsibility
ER	Environmental Report
EU	European Union
FEB RAS	Far Eastern Branch of the Russian Academy of Sciences
GARP	Global Atmospheric Research Programme
GCMs	General Circulation Models
GGI	Russian State Hydrological Institute
GGO	Voeikov Main Geophysical Observatory
GHG	Greenhouse gas
GMT	Global Mean Temperature
IGY	International Geophysical Year
IPCC	Intergovernmental Panel on Climate Change
LAM	Laboratory for Monitoring the Environment and the Climate (Moscow)
LULUCF	Land use, land use change and forestry
NGO	Non-governmental organisation
NWP	Numerical Weather Prediction
OSA	Open Society Archives
PICES	North Pacific Marine Science Organization
POLEX	Polar Experiment (part of GARP)

RFE/RL	Radio Free Europe/Radio Liberty
	Roshydromet (or Gidromet) Federal Service for Hydrometeorology and Environmental Monitoring
RSFSR	Russian Soviet Federative Socialist Republic
SR	Sustainability Report
UN	United Nations
UNESCO	United Nations Educational, Scientific and Cultural Organization
UNFCCC	United Nations Framework Convention on Climate Change
US/A	United States/of America
USSR	Union of Soviet and Socialist Republics
WMO	World Meteorological Organisation
WTO	World Trade Organisation
WWF	World Wildlife Fund

Note on transliteration

In this book, we use a Library of Congress transliteration system for Russian language. Well-known names appear in their most common transliterated form (for example, Bedritsky instead of Bedritskii).

Preface

The book *Climate Change Discourse in Russia: Past and Present* explores the relationship between the science, politics and policy of climate change in the post-Soviet world by considering the historical development of the concept of climate change as a scientific thesis, environmental problem and policy issue. The book's chapters complement work on the history of the discursive political construction of climate change in the West by examining a highly contrasting (but intimately related) cultural context.

The early drafts of all contributions to this volume were first presented at the workshop *Soviet and Post-Soviet Imaginings of Climate* which took place in 2017 at King's College, London. The event was organised by Nottingham Trent University (Department of Politics and IR) with support from King's College, (Russia Institute) and had the financial support of the British Association of Slavic and East European Studies (BASEES) and the Academic Association for Contemporary European Studies (UACES). The event presented an excellent opportunity for researchers working on the social aspects of climate change in the post-Soviet space to advance the discussion of this highly debated environmental problem in the region.

By the end of the workshop, it had become apparent to all its participants that there is an urgent need to continue enhancing our knowledge of the Soviet influence on contemporary understandings of climate, climate change and environmentalism as well as modern Russia's official and popular discourses. Furthermore, it was concluded that we need to try to uncover how Russia's past has affected and explain its complex present with regard to the climate change narratives in the country. It is with these deliberations in mind that we have come to the idea of the present volume, which has incorporated the advanced, peer-reviewed and updated versions of the most suitable workshop papers.

It is our hope that the book will not be treated as a conclusion to this complex and acute discussion, but serve as another piece of dialogue in an ongoing academic exploration. As the world begins to suffer the extreme consequences of anthropogenic climate change, finding adequate solutions to global environmental problems necessitates the participation of all countries. The role of Russia may be a central one in determining the nature and success of these global solutions and it is, therefore, increasingly important to understand climate change discourse in the region. Insights gained in this area may also be illuminating for examining environmental discourses in other resource-rich regions of the world with alternative economic and political experiences to that of the West (e.g. China, the Middle East).

Marianna Poberezhskaya
Teresa Ashe

1 Introduction

The importance of the Soviet experience

Teresa Ashe

Introduction

Climate change has been a scientific issue for many decades and has, since the 1990s, dominated environmental politics at both domestic and international levels. Scholarship on the social history and cultural dimensions of 'climate change' as a concept has, over recent decades, made clear how and why our understanding of this phenomenon has developed the way it has. Such research allows for greater nuance and understanding of the many ways the topic is discussed and engaged with in different social spheres. However, the understanding that we have of climate change and its history is still, in the Anglophone literature, predominantly a story about American science and politics shaping global understandings of the phenomenon.

Despite the European origins of the theory of anthropogenic global warming, it was in America that the idea took scientific root. American research drove the development of climate science. American domestic politics shaped political understandings of the environmental problem. American negotiating strongly influenced the character of the United Nations Framework Convention on Climate Change (UNFCCC) and its evolution. The central role of American scientists and political actors in shaping the way climate change is understood today cannot be denied. And yet, important as this role is, there are other cultures of climate change which exhibit similar discursive vitality in shaping shared global understandings of the issue.

This book advocates the position that one of the most important and under-studied cultures of climate change is that of Russia, both in the Soviet and post-Soviet periods. As a key belligerent in the Second World War and a superpower of the Cold War, it is apparent that the technological and scientific pursuits of this nation, at the very least indirectly inspired and shaped the scientific organisation of the Western powers. Yet, more directly, Soviet scientists in the twentieth century were at the cutting edge of meteorological study and vastly influential in the global scientific community. The political actions of Soviet nations were fundamental in shaping scientific and environmental organisations like the International Geophysical Year (IGY) in 1957/8 and the Intergovernmental Panel on Climate Change (IPCC) created in the 1980s. Post-Soviet Russia also had an important role to play in the politics of climate change internationally. Without US ratification of the Kyoto Protocol, it was Russian participation that carried the Protocol over the threshold for effectiveness and allowed it to come into force. Russia currently holds a special place in international climate change negotiations, being not only a member of the BRICS block (Brazil, Russia, India, China and South Africa) of rapidly developing countries, but also one of the few nations in the world that has claimed with any plausibility that it might benefit from a warmer climate, at least in the short-term. It is also one of the nations about which the Anglophone literature has the least to say.

This book seeks to remedy this relative silence by drawing together contributions from a number of scholars researching in this area. Their approaches are by no means homogeneous and their areas of interest vary widely, but what unites them is their interest in Soviet and post-Soviet scientific, political and policy discourses of climate change and the environment. In this sense, the book is broadly rooted in the Discourse Analytical tradition of Michel Foucault, which traces the origins and power structures underpinning the social construction of a topic, deepening understanding of how a subject changes over time and of how those changes structure attendant social and political possibilities.

To contextualise and introduce the chapters in this volume, this introduction first outlines a short history of the scientific and political conceptualisation of climate change in the West, making clear how the centrality of the American experience requires a complementary appreciation of the Soviet experience. A brief literature review of the Anglophone literature on Soviet and post-Soviet climate change discourse will be presented in order to show that this is an under-studied area, but one in which important scholarship is currently being conducted. It will then outline the collection of chapters in this book and how severally and collectively they contribute to mitigating the deficiencies of the Anglophone literature on this topic.

A short history of climate change

Climate change as a scientific idea is usually dated back to a paper by Svante Arrhenius in 1896, in which he postulated that, because humans were adding quantities of carbon dioxide to the atmosphere through the burning of fossil fuels (something on which his colleague Arvid Högbom was working), this would contribute a warming influence that could change the climate of the planet (Arrhenius, 1896). Arrhenius' rough calculation was that a doubling of CO_2 in the atmosphere would lead to a 2-degree warming of Global Mean Temperature (GMT). However, he was in no way suggesting that this should be regarded as problematic. In fact, the debate to which this paper was a contribution assessed the possibility, not of a warming planet, but of future ice ages.

The idea of climatic change had been difficult to countenance at the beginning of the nineteenth century, when 'climate' was essentially defined as an averaging of recorded weather patterns over a given period of time. However, in 1837, Louis Aggasiz proposed that many European landscapes could be better explained by the idea of historic ice ages than by the commonly held diluvial theories of nineteenth-century geology, which owed much of their explanatory power to biblical authority. With this idea of historic ice ages came the threat of potential future ice ages, and the scientific community began to propose theories that might explain how global temperature could change on such a dramatic level.

Conceptualisations of how the planet worked were simplistic enough at this time that it was presumed there would be one or two factors that might account for changes to the climate, and, of the many theories proposed, some suggested that a return to ice age climates might be likely. There was concern that if climates could change, then there was reason to fear a future ice age. Hence, by the time Arrhenius made his famous contribution to the debate, the idea that changes in atmospheric gases might account for planetary temperature changes was a fairly comforting notion: if human activities *were* to have such an effect, it would be a *counterweight* to any factors tending towards a potential ice age.

By the 1930s, there were scholars arguing that anthropogenic global warming was already discernible and should be taken seriously as a scientific phenomenon (Callendar, 1937). However, the majority of scientists at this time thought, that adding more CO_2 to the atmosphere would be unlikely to have any additional affect. Studies suggested that under normal conditions the spectrum at which CO_2 is able to block additional infrared radiation would already be saturated. Additionally, were this to prove untrue, there was in any case little reason for concern about anthropogenic warming, because *if* climate change of this kind were to happen, it was thought, it would be a positive thing, protecting humanity from the threat of ice ages.

The scientific and social shifts that allowed climate change, as a scientific theory, to garner more interest, came about in the context of the Second World War. It was during the Second World War that routine data collection and research at altitude made the scientific community aware that additional atmospheric CO_2 *was* capable of blocking additional infrared radiation in the upper atmosphere and therefore that anthropogenic increases in CO_2 could be having an effect on the climate (Plass, 1956). It was also Second World War technological developments (the computer and the nuclear bomb, particularly) that facilitated post-war research into planetary systems, leading to the institutionalisation of climate change as a viable scientific topic of study. Second World War technological advancements and Cold War motivations made understanding the atmospheric system an important goal for post-war America, and it was here that climate change research in both oceanography and meteorology took root.

Oceanography was important to the study of climate change because, after it was accepted that the effect of CO_2 at altitude *could* affect the heat balance of the Earth, the next best reason for scientists to dismiss the idea was that the oceans would probably absorb any extra CO_2 humanity might emit. This was a plausible assumption until a paper by Roger Revelle and Hans Suess, which studied the way ocean chemistry responded to nuclear fallout, was published in 1957. This paper queried the assumption that the oceans would prevent any anthropogenic warming effect by absorbing excess CO_2 and offered calculations of ocean chemistry that suggested that initial absorption might actually be followed by a relatively immediate re-release of significant quantities of CO_2. The idea that humanity might be affecting the planet in this way provoked surprise and curiosity within the oceanographic community, but it did not cause marked alarm or facilitate a narrative that would be appealing to funders of science. Revelle advocated for a study of baseline CO_2 in the atmosphere, which, through the institutional funds of the International Geophysical Year in 1957/8, allowed Charles Keeling to begin the research project that has since given us the famous Keeling Curve (Weart, 2008). This curve shows baseline CO_2 in the atmosphere rising over time in a seasonal 'saw-tooth' pattern that confirms the idea that oceans are not absorbing excess CO_2. Thus, an important, but modest, programme of research was established in oceanography that was able to furnish supportive empirical data on CO_2 and confirm that scientific understandings of atmospheric CO_2 needed re-evaluation. However, there was still a prevailing sense in the 1950s and 60s that any global warming effect from CO_2 increases would be a welcome counterbalance to potential global cooling.

In meteorology, parallel scientific developments in the 1950s and 60s furnished another re-evaluation of the idea that human activity might affect the climate, but the context was quite different. Rather than provoking curiosity, the idea was instead inspiring. Rather than the securing of a modest empirical research programme, the notion of climatic change formed part of the inspiration for an ambitious, theoretical research programme that would take advantage of cutting edge technologies to address some of the key social issues of the period. John von Neumann, who had been involved in the development of both the computer and the nuclear bomb, attempted to harness the power of the former in the service of numerical weather prediction (NWP), which meant taking readings of meteorological variables and, by solving equations based on this initial state, project what future meteorological states these variables would pass through over time. In essence, NWP was an attempt to forecast the weather, not from macro-level data such as mapping an incoming warm or cold front, but from a set of equations that would tell the researcher how weather would progress, given only an initial set of variables. This had been attempted by meteorologists before but had always previously failed due to lack of computational power. Now, with the computer to run calculations, the first successful hindcast (of weather events that had already happened and so could be checked) was achieved in 1950.

One may ask why, given that this research had little to do with atmospheric CO_2 or potential changes to climate, it should loom large in the history of climate science. One might also ask in what ways it could be said to be 'inspired' by climate change. The answers to these questions lie in the specifically Cold War context of the research project and the funding environment of the time. Interest in reliable NWP was desirable not solely for the obvious civilian and military benefits of knowing what meteorological conditions to anticipate at a given date in the future, but also because, if successful, it represented a first step along a pathway to understanding the weather well enough that it could be controlled and manipulated. Anthropogenic global warming was not a central focus or concern of this research programme, yet the possibility of climate change here represented, not a worrying environmental threat, but an inspiring hope of climate system knowledge with military applications. The idea of anthropogenic global warming incentivised pursuit of weather studies: if it could be understood whether human society was *inadvertently* changing the weather through the emission of CO_2, then it could better be understood how to *deliberately* change the weather to favour the American and disadvantage the Soviet, economic blocks.

By the 1960s, the modelling undertaken to understand weather events had developed to allow for the creation of a General Circulation Model (GCM) of the atmosphere. The first of these was attributed to Norman Phillips in 1955 and, though it modelled only a single column of atmosphere, it was followed by models that undertook to represent the entire globe (Weart, 2003: 59). These models did not just try to predict weather from a given set of data, but aimed to create climate models that mimicked the observed weather patterns of an earth typed planet. These facilitated far greater understanding of the planetary system and allowed for the prediction of likely weather patterns over far greater time scales than the weather models had permitted. In 1975, Syukuro Manabe and Richard T. Wetherald tested Arrhenius's old hypothesis that a doubling of CO_2 in the atmosphere would lead to a 2-degree warming of the planet. Their interest was in the heat balance of the earth and not directly in the risks of CO_2 emission, but their findings confirmed Arrhenius's basic theory and made CO_2 warming a viable research topic in climatology (Manabe and Wetherald, 1975).

Thus, the idea of anthropogenic climate change was, by the mid-1970s, empirically plausible, with evidence that baseline atmospheric CO_2 was rising, and theoretically plausible, as GCMs confirmed what had previously been 'back of the envelope' calculations of the effects of doubling CO_2. In both oceanography and meteorology insights generated by the investigation of other topics had made the notion of anthropogenic global warming viable. The Cold War context here was vital. After the Second World War geo-scientific knowledge was amply appreciated by belligerent states and mounting Cold War tensions did little to devalue this type of knowledge. Both Roger Revelle and John von Neumann were working on projects related to Second World War technologies and Cold War problematics when they made their contributions to climate change science. Revelle was studying the role of oceanic chemistry with a view to better appreciating nuclear fallout (so that Soviet testing could be better apprehended), and von Neumann was using the computer in a project with consciously recognised military applications (predicting the conditions in which troops would be fighting and ultimately perhaps controlling those conditions).

On the other hand, cross-border international cooperation in the cause of science was often viewed as a palliative to such Cold War tensions and thus scientific activity was encouraged on both sides of the Iron Curtain. The International Geophysical Year (IGY) in 1957/8 represented this pacific discourse, encouraging nations to work together in the pursuit of geophysical knowledge. Yet the IGY also provided opportunities for the superpowers to assess the extent of each other's expertise and technology, not to mention the potential to demonstrate the relative superiority of the First- or Second-World approach to science. As such, scientific prowess was viewed as emphasising the superiority of the way of life to which each superpower was ideologically committed.

The military context of the Cold War shaped the topics that were of interest to American scientists, the tools with which they tackled these topics, the attitude their governments took to scientific activity and also the funding regimes that made such research possible. The Cold War American funding context in which the two research programmes took route was highly contoured by the government's hope for viable military knowledge and national prestige. While von Neumann and his NWP research community firmly situated their work on weather prediction and manipulation as a potential step towards Cold War weather weaponry, the oceanographic research community was much less inclined to invite the oversight of the military funding bodies and preferred to emphasise that their work was 'basic science'.

The way climate research was framed by meteorologists, therefore, was explicit that anthropogenic changes to weather might be possible, but this was not at all presented as a cause for concern. The idea that human beings could modify planetary weather systems in their own interests was an attractive and powerful idea for Cold War 'hawks' like Von Neumann and his military backers. The oceanographic researchers were much less keen to flag the potential for weather weaponry, but were equally far from framing CO_2 warming as a potential environmental problem. It was simply a curious feature of the oceanic and atmospheric system, which warranted attention as part of a basic research agenda. Thus, climate change had not, at this time, acquired the characterisation of being environmental or problematic in the way we would recognise today.

The shift from climate change as part of a basic research or Cold War weaponry agenda to climate change as an environmental problem was a result of both social and scientific changes in the 1960s and 1970s. GCMs were shifting perceptions of the way planetary dynamics worked and creating a starkly different understanding of the natural world to the one that had characterised the previous century's search for the one or two factors that might dictate changes in global temperature. In the nineteenth century, the natural world had been imagined as 'mechanistic', but this view of the world was becoming untenable in the light of new scientific techniques. Prior to the advent of computers, study of the natural world had been better able to cope with smooth curves and predictable patterns and so technical capabilities had obscured elements of natural systems that were not amenable to the computational capacities of the time. This enforced the worldview that nature was essentially regular and ordered. Computer modelling of weather and climate models led to understandings of the natural world as being far more intricate than was previously appreciated. Increasing evidence from computational investigation showed that planetary systems were actually inherently complex, in the sense that smooth curves and patterns were not the norm. This insight began to raise concerns about the instability and fragility of planetary conditions that had once been imagined as robust and unassailable.

As such, the study of human activity affecting global weather was initially conducted in a spirit of optimistic prometheanism, that America might one day know the weather system well enough to manipulate it for Cold War gains. Yet the research that hoped to facilitate this was uncovering a much more complex picture of the cause and effects of weather and climatic changes. This meant that weather modification projects, despite being relatively successful, ceased to be viewed as viable. This was not because they did not deliver the benefits they promised, but because the complex nature of the climatic system made it difficult to isolate the desired and predicted effects from a plethora of other consequences that might or might not be attributable to the intervention. There was thus a shift away from seeing the planetary system as robust, stable and amenable to specific and measurable interventions. Instead, it began to appear as complex, vulnerable and 'unpredictable in principle'. This made the idea of weather modification seem less a reliable tool in the arsenal of Cold War weaponry and more a risk that might have consequences which would far outreach those originally intended.

This shift of perspective from a promethean desire to control nature, to an environmental concern that nature might be adversely affected by human activity, shifted the way research into climate change was undertaken. Rather than being framed as a precursor to weather modification research, climate change began to be viewed as a potential environmental problem, but not without reserve. While it was increasingly clear that atmospheric gases could change the temperature of the planet, this was just one of the many variables considered and while anthropogenic CO_2 concentrations might be warming the planet, it was thought that other factors were still making global cooling a more likely future outcome. It was not until the 1980s that research into the magnitude of global cooling trends and more evidence linking historic GHG levels to historic temperature changes led US scientists to place climate change on the policy agenda, and it was soon after recognised as a global problem.

The cultural development of the idea of 'climate change' from obscure scientific theory, to viable research problem, potential weapon, environmental concern and, eventually, environmental policy problem, cannot be read as an inevitable or prefigured trajectory. Reliant as it is on technological and social developments that shaped the timing and progress of the science and the social context in which it was understood, funded and interpreted, this history is situated and contingent. One of its key features is the Cold War context, which played an important role in shaping how the western apprehension of climate change developed. Yet, the Anglophone literature has little to say on the parallel developments in the science of climate and the politics of environment in the other superpower. This lacuna is problematic given the centrality of the funding rhetoric of the 1950s: that weather modification research was doubtless taking place in the Soviet Union and that America must fund climatic research in order to keep up with it.

The sharing of science under the Cold War framework meant that Soviet scientists contributed greatly to the developments in the geosciences after the Second World War, but the political sensitivity of this area obscures the extent of collaboration and of isolation. Very different political and scientific cultures also raise questions about the nature of environmental thought in the USSR and how it developed, which research areas were pursued, what results were produced and how was such scholarship presented so as to attract what sort of funding and support. This is where the contributions of this volume can begin to illuminate the obscure elements of this history of climate change and make their contributions to the budding literature on Russian contributions to the science and politics of climate change.

Soviet and post-Soviet discourses of climate change: a review of the literature

This book adopts a discourse analytic approach to the politics of climate change. This approach is interested in how we conceptualise and discuss environmental ideas and how those conceptualisations structure the responses we make to environmental problems. Discourse analytic approaches focus on patterns of language and representation in order to trace how a conceptualisation changes over time, over space or between different discussants. The study of environmental politics in the West has benefited significantly from the discourse analytic approach to international politics, with scholarship examining how the framing and conceptualisation of environmental issues offers an important supplement to the traditional international relations analysis focus on interests and power.

Environmental Discourse analysis, usually drawing on the Foucauldian notion of genealogy, shows that discourse shapes perceptions of interest and structures different kinds of power. The power of scientific expertise is of particular interests in environmental discourse. The study of how certain 'ways of knowing' can hold power has been successfully deployed by Karen Litfin (1994) in the study of ozone depletion, Maarten Hajer (1995) in the study of acid rain in Europe and John Dryzek (2005), who instead catalogued different ways of discursively constructing environmental issues generally. My own research (Ashe, 2011; Ashe, forthcoming) has sought to do similar work for the topic of climate change.

The aim of this book is to draw together the contributions of researchers working on different periods and in different disciplines but who are each contributing to the study of climate change discourse in Russia. Through this synthesis, the aims of environmental discourse analysis will be reached while also allowing a greater appreciation of the different disciplines and approaches that contribute to a better understanding of climate change discourse.

In turning to the topic of Russian environmental discourse, there is an emerging literature exploring various aspects of environmental history and awareness in the Soviet and post-Soviet periods.

Douglas Weiner's (1999) study of environmental movements in the Soviet Union helps make clear how environmental ideas thrived, even in a society restrictive of critical thought. He argues that there is no singular explanation for the relative licence given to this dissenting social movement, but mentions that the scientists involved were not seen as a viable 'political' threat by the authorities, they positioned themselves as 'naïve' nature lovers, rather than dissidents. Weiner makes clear that the nature protection movement was genuinely, rather than cynically, patriotic and had no intention of overthrowing the Soviet system, which it often felt more likely to make the kind of grand nature protection policy it advocated than a more democratic system might. The movement also had high- and middle-ranking patrons which helped protect it and made sure it never positioned itself as too antagonistic to the existing power structures. Weiner makes clear that the movement was by no means unitary and reflects on the diversity of meanings the idea of 'nature protection' had within the Soviet Union. The focus of Weiner's study is on nature advocacy as a form of social identity created in tension with, but not necessarily antagonism with, the Stalinist state. He also makes clear that the 'nature protection movement' he studies is multifaceted:

> Complicating this task is that nature protection provided the symbols and rhetoric around which more than one distinct, autonomous subculture was organized in Russia. Consequently, this is a study of how 'nature protection' as an aesthetic, moral, and scientific concern and as a source of symbols and rhetoric was used creatively by Soviet people to forge or affirm various independent, unofficial, but defining social identities for themselves.
>
> (Weiner, 1999: 20)

Paul Josephson et al. (2013) offer *An Environmental History of Russia*, which explains the natural resource management policies of the Soviet Union. This account focuses more on the political and economic factors shaping how Soviet elites tried to manage the vast and diverse Russian landscapes. They highlight similarities between the development of environmental thought in Russia and in the West, but note the lack of a developed civil society in the authoritarian Soviet regime as a key difference between the two, explaining why a political environmental social movement did not emerge in Russia in the same way. Henry (2010) in her book *Red to Green: Environmental Activism in Post-Soviet Russia* considers post-soviet environmental efforts as a case study for understanding the possibilities of citizen mobilization and its effect in Russia.

In addition to the study of environmental social movements and nature policy, there is literature considering the emergence of Russian environmental knowledge. Jonathan Oldfield and Denis Shaw draw on the work of David Livingstone's (1953) *Putting Science in its Place*, which considers the 'geographies of scientific knowledge', examining how science can have roots in particular cultures and places. This approach shapes Oldfield and Shaw's (2016) *The Development of Russian Environmental Thought*, studying the emergence of the discipline of Geography in Russia and its advances during the Soviet period while charting emerging environmental ideas.

In the study of Russian environmental political discourse, there are scholars who have consciously adopted Foucauldian approaches, such as Geir Hønneland, whose book *Russia and the West: Environmental co-operation and conflict*, published in 2003, studied the way that Russia interacts with the West in the European Arctic. Hønneland's case studies in this volume are 'marine living resources', 'nuclear safety' and 'industrial pollution' (Hønneland, 2003: 2). A more recent book, Anna Korppoo, Nina Tynkkynen and Geir Hønneland's (2015) *Russia and the Politics of International Environmental Regimes*, takes the same environmental discourse analytic approach to climate policy, water protection and fisheries management, exploring how far environmental encounters can be understood as foreign policy. Hønneland (2016) continues discourse analysis of the Arctic, but this time with a focus on the identity narratives discernible in Russian representations of the post-2007 'scramble for the Arctic'. Hence, there is an emerging body of literature on environmental discourses in Russian international politics.

In addition, there is a growing literature more specifically on Russia's politics of climate change. Elana Wilson Rowe's (2013) *Russian Climate Politics: When Science Meets Policy*, draws on the work of Sheila Jasanoff and Brian Wynne in studying how knowledge translates and transfers to new institutional and geographical settings. Yet, like environmental discourse analysts, it focuses on how scientific knowledge is situated socially and how this affects environmental politics. It thus has strong affinities with Foucauldian focusses on how knowledge holds power.

Another literature we might expect to have studied Russian discourses of climate change is the literature on climate change communication and media depictions of climate change. However, this has not provided much coverage of Russia. Wilson Rowe (2009) looked at the reporting of climate change in *Rossiskaia Gazeta* in order to assess how climate change is framed in Russia and Tynkkynen (2010) also considered the framing of climate change as a policy issue. There are comparative studies that look at climate change in the print media of various countries, in which Russia has been included as one of the nations considered. Painter's (2010) study of international media representations of climate change and Schmidt et al.'s (2013) study of media attention to climate change in newspapers both include Russia.

In 2016, Marianna Poberezhskaya's *Communicating Climate Change in Russia; State and Propaganda* looked in more depth at representations of climate change in the Russian print media. Exploring how the Russian media has developed through regime change and a shift to a market economy, how it has reported climate change and how state policy has changed alongside it, Poberezhskaya is able to consider how far state policy influences climate change communication in Russia and offer a more thorough look at the climate change communication in this region.

Thus, there are literatures emerging on environmental movements, environmental science and environmental policy discourse in Russia. There are also studies of the policy discourses and media representations of climate change. Yet, in none of these areas is there abundant scholarship, making clear that this is an area ripe for further research. There is an emerging picture of the complexity and breadth of environmental and climate discourse in Soviet and Post-Soviet Russia, raising intriguing questions for further research.

Overview

As a scholarly volume, this book is rooted in the orthodox understanding of the science and politics of climate change. However, the editors believe that a deeper understanding of the politics of climate change can be achieved by appreciation of the various cultures of climate change that are held by different nations, even heterodox discourses such as climate change scepticism or denial. Russia, as a significant nation in climate change politics, has its own unique scientific, political and policy experience of climate change and this book aims to illuminate that unique experience. Such an endeavour is always premised on a fruitful tension between its ontological foundation in scientific orthodoxy, and its desire to make meaningful and accessible a greater understanding of the diverse and sometimes heterodox discourses of the environment. Without belief that the climate is changing, few scholars would think it worthy of study and yet without the ability to suspend that belief, few scholarships of climate cultures would be worthy of the name. In the case of Russia, an appreciation of the foundations of both climate change policy advocacy and climate change scepticism are equally necessary.

The first two chapters look at the Soviet understanding of climate and environment, while the following four chapters look at contemporary discourses of climate change in the national print media and elite political discourse (Chapter 4), national Russian print media discourse (Chapter 5); local print media and scientific discourse in Russia's Far East (Chapter 6) and the discourse of Russia's business community (Chapter 7).

Chapter 2, by Katja Doose and Jonathan Oldfield, examines understandings of natural and anthropogenic climate change in the Soviet Union during the period from the 1960s through to the early 1990s. This chapter makes clear the contributions made by Soviet scientists to climate science and international governance of science and environment. It also gives a sense of the orthodox scientific discourses that contributed to institutionalised understandings of science in the Soviet Union.

This contrasts with the work of Anna Mazanik, who, in Chapter 3, looks at media discourses of the environment and of climate change specifically from the late 1960s to the early 1980s. She looks both at the mainstream Soviet press and at the uncensored narratives communicated in samizdat. Mazanik makes clear that environmental issues were discussed in the mainstream media and that scientists were a key vehicle for communicating about them, particularly in the case of climate change. Yet, in the samizdat collections, she finds a relative dearth of environmental coverage, suggesting that, while sometimes of interest, environmental issues did not form an important strand of dissident narratives and groups in the 1950s to 1980s.

Veli-Pekka Tynkkynen's Chapter 4 continues the focus on discourses of the environment by considering the post-Soviet environmental narrative under Vladimir Putin's presidency. He argues that, while Medvedev courted Europe with promises of environmentally responsible energy, a changing energy market and increasing international tensions made it less important for Putin to maintain this image. It was thus more tempting to position Russia as an energy superpower, blending political discourses and a national identity bound up with 'hydrocarbon culture', so that climate change denial became a natural corollary to this. Both climate change denial and a hydrocarbon culture, as exemplified in Gazprom's promotional materials in the early 2010s, are then explored.

Chapter 5, by Marianna Poberezhskaya, looks at Russia's official climate change discourse through the lens of the mainstream print media in Russia. This allows her to examine not only the elite and media discourses on climate change, but also the changing role of the mainstream media in Russia. She tracks the development of Russian climate change discourse in Russian national newspaper, *Izvestiia*, from 1992 to 2012. Here she makes clear that media representations of climate change have closely tracked policy positioning and also notes a growing unwillingness to present climate change as anthropogenic.

Chapter 6 looks at more geographically marginal contemporary discourses of climate change by examining climate change in the Russian Far East. In this chapter, Benjamin Beuerle looks at local print media and interviews with scientific and environmental elites in Vladivostok, Khabarovsk and Petropavlovsk-Kamchatsky. His chapter looks at the regional impacts of climate change, regional engagement with the topic, how the Asia-Pacific influences climate attitudes in the area, the importance of climate scepticism in the region and the development of renewed forms of engagement with climate change that may help overcome this scepticism.

In Chapter 7, Ellie Martus looks at industry narratives of climate change in Russia and their contribution to wider climate change discourse. Her focus is the oil and gas companies, which are responsible for 82.16% of Russia's total GHG emissions (excluding LULUCF). By examining companies' Environmental, Sustainable or Annual Reports and/or company websites, Martus examines the way that the energy industry in Russia present their attitudes to climate change and assesses how far being perceived as a responsible environmental actor is a priority to such companies. She finds reference to environmentally responsible (generally energy saving) practices in a number of cases, but little sense of climate change as a business risk or engagement with international climate policy. She also considers the position of the energy industry narrative in comparison to that of the Russian mining industries.

Concluding remarks

This introduction has aimed to make clear why the Russian cultural experience of climate change is an important one. The centrality of the Cold War context in shaping development of climate change science and politics in the US raises questions about the Soviet experience, some of which are answered in Chapters 2 and 3 of this volume. Yet, the Soviet context also helps us understand that the Western understanding of climate change is contingent on a particular scientific and political experience, which may not be shared by other regions. The chapters in this volume build on the beginnings of an important literature in the climate change cultures of Russia to offer greater insight into a region that mirrors the experience of environmental issues in Western Europe and the US, yet creates its own unique culture of climate change that in turn shapes how policy makers, scientists, media, industry and the public conceptualise this issue.

Chapters 4, 5, 6 and 7 in this volume explore how climate change is being imagined, discussed, constructed and projected in different contexts and by different actors in Russian society. It is clear that the 'hydrocarbon culture' described in Chapter 4 contextualises the national (Chapter 5) and the local (Chapter 6) media discourses, as well as the business discourses (Chapter 7) explored in the post-Soviet part of the book. Yet this volume makes clear that the discourse of climate change in Russia is by no mean fixed. It has changed historically and will doubtless develop in the future. There are moments and voices of support for climate action as well as factors militating against it.

It is hoped that this volume develops understanding between those from different environmental cultures and promotes dialogue that recognises and draws on the diversity of understandings of environmental issues. It is clear that Russia has contributed greatly to the history of climate change science and politics, yet it also has strong tendencies to maintain 'business as usual' policies as part of its 'hydro-carbon culture'. A greater appreciation of the choices Russia faces and the cultures that structure its decisions will doubt-less be a great asset in understanding the future of global efforts to mitigate and adapt to anthropogenic climate change.

References

Arrhenius, S. (1896) 'On the influence of carbonic acid in the air upon the tempera-ture of the ground', *Philosophical Magazine and Journal of Science*, vol 5, no 41, pp. 237–276.

Ashe, T. (2011) *The Politics of Climate Change: Power and Knowledge in Environ-mental Politics* (Unpublished PhD Thesis, Birkbeck College, University of London).

Ashe, T. (Forthcoming) *Climate Change: Discourses of Making and Unmaking.*

Callendar, G.S. (1937) 'The artificial production of carbon dioxide and its infl-uence on temperature', *Quarterly Journal Royal Meteorological Society*, vol 64, pp. 223–240.

Dryzek, J. (2005) *The Politics of the Earth: Environmental Discourses*. Oxford Uni-versity Press, New York.

Hajer, M. (1995) *The Politics of Environmental Discourse: Ecological Moderniza-tion and the Policy Process*. Oxford University Press, Oxford.

Henry, L.A. (2010) *Red to Green: Environmental Activism in Post-Soviet Russia*. Cornell University Press, Ithica.

Hønneland, G. (2003) *Russia and the West: Environmental Co-Operation and Conflict*. Routledge, Abington.

Hønneland, G. (2016) *Russia and the Arctic: Environment, Identity and Foreign Policy*. I.B.Tauris, London.

Josephson, P., Dronin, N., Mnatsakanian, R., Cherp, A., Efremenko, D. and Larin, V. (2013) *An Environmental History of Russia*. Cambridge University Press, Cambridge.

Korppoo, A., Tynkkynen, N. and Hønneland, G. (2015) *Russian and the Politics of International Regimes: Environmental Encounters or Foreign Policy?* Edward Elgar, Cheltenham.

Litfin, K. (1994) *Ozone Discourses: Science and Politics in Global Environmental Cooperation*. Columbia University Press, Chichester.

Livingstone, D. N. (2003) *Putting Science in its Place: Geographies of Scientific Knowledge*. University of Chicago Press Ltd., London.

Manabe, S. and Wetherald, R.T. (1975) 'The effects of doubling the CO2 concentra-tions on the climate of a general circulation model', *Journal of the Atmospheric Sciences*, vol 32, no 1, pp. 3–15.

Oldfield, J.D. and Shaw, D.J.B. (2016) *The Development of Russian Environmental Thought: Scientific and Geographical Perspectives on the Natural Environment*. Abingdon, Oxford.

Painter, J. (2010) *Summoned by Science: Reporting Climate Change at Copenhagen and Beyond*. Reuters Institute for the Study of Journalism, Oxford.

Plass, G. (1956) 'Carbon dioxide and climate', *Scientific American*, vol 201, pp. 41–47.

Poberezhskaya, M. (2016) *Communicating Climate Change in Russia: State and Propaganda*. Routledge, Abingdon.

Revelle, R. and Suess, H.E. (1957) 'Carbon dioxide exchange between atmosphere and ocean and the question of an increase of atmospheric CO2 during the past decades', *Tellus*, vol 9, no 1, pp. 18–27.

Schmidt, A., Ivanova, A. and Schaefer, M.S. (2013) 'Media attention for climate change around the world: a comparative analysis of newspaper coverage in 27 countries', *Global Environmental Change*, vol 23, pp. 1233–1248.

Tynkkynen, N. (2010) 'A great ecological power in global climate policy? framing climate change as a policy problem in Russian public discussion', *Environmental Politics*, vol 19, pp. 179–195.

Weart, S. (2003) *The Discovery of Global Warming*. Harvard University Press, London.

Weart, S. (2008) *Money for Keeling: Monitoring CO2 Levels*, www.aip.org/history/climate/Kfunds.htm.

Weiner, D.R. (1999) *A Little Corner of Freedom: Russian Nature Protection form Stalin to Grobachev*. University of California Press, Berkeley.

Wilson Rowe, E. (2009) 'Who is to blame? Agency, causality, responsibility and the role of experts in Russian framings of global climate change', *Europe-Asia Studies*, vol 61, no 4, pp. 593–619.

Wilson Rowe, E. (2013) *Russian Climate Politics: When Science Meets Policy*. PalgraveMacmillan, Basingstoke/New York.

2 Natural and anthropogenic climate change understanding in the Soviet Union, 1960s–1980s

Katja Doose and Jonathan Oldfield

Introduction[1]

Soviet climatologists and cognate scientists were heavily involved in the developing discussion around anthropogenic climate change that emerged on the global stage during the second half of the twentieth century. Through a combination of domestic endeavour and international exchange, their science was at the forefront of work related to understanding climate change as well as initiatives to anticipate and predict warming trends during the twenty-first century. For example, Soviet climate scientists formed an integral part of the activities of the 1957–1958 International Geophysical Year, which was a key milestone in our growing awareness of large-scale physical processes, including the climate (see Collis and Dodds, 2008). Soviet scientists were also influential in the lead-up to the first report of the Intergovernmental Panel on Climate Change (IPCC) in 1990. Nevertheless, the nature and extent of their contributions to this debate are generally underplayed within dominant English-language overviews, which typically stress the slow, meandering accretion of relevant knowledge evident within the European and North American contexts from the mid-nineteenth century onwards (e.g. Weart, 2003).[2] Where reference is made to Soviet science, it is often linked to the activities of a handful of dominant figures who achieved prominence internationally through their innovative research in areas such as quantitative climatology (M.I. Budyko [1920–2001]), remote-sensing (K.Ya. Kondratyev [1920–2006]) and related areas, in addition to their work with international bodies such as the World Meteorological Organisation (WMO) (e.g. G.I. Marchuk [1925–2013]) and the Intergovernmental Panel on Climate Change (IPCC) (e.g. Yu.A. Izrael [1930–2014]). The contribution of these individuals was substantial and yet a focus on their work shifts attention away from the developed domestic discussion around the broader question of climate change as well as from other actors involved in these debates. Soviet climate science is given more visibility in the

English-language literature with respect to a series of pronouncements from Soviet scientists during the late 1980s and early 1990s that emphasised the benefits of global warming, particularly linked to anticipated increases in levels of agricultural productivity (e.g. Miller and Pearce, 1989: 24). These assertions need placing within the broader sweep of Soviet work related to anthropogenic climate change since they fail to capture effectively the overall tenor of Soviet activity in this area. Broadly speaking, focussed work concerning Soviet climate science and, in particular climate change, has a number of potential uses. First, it will help to deepen our appreciation of the Soviet institutional framework within which climate science and related areas of science operated, providing a much-needed insight into the work of the country's influential hydrometeorological services. Second, it will assist in advancing understanding of the various ways in which Soviet science was communicated on the international level during the Cold War period. Third, it will provide insight into the way in which Soviet climatologists and cognate scientists conceptualised the interaction between climate and society.

This chapter offers an initial foray into aspects of these three areas of interest by highlighting some key features of the Soviet Union's engagement with the scientific discussions around climate change from the 1960s onwards. In order to do this, the chapter is divided into the following main sections. First, we briefly highlight the longstanding nature of Soviet interest in meteorological and climatological issues and associated institutional initiatives underpinning this activity. Second, we reflect on the emergence of climate change science within the Soviet Union, with specific reference to the work of individuals linked to key institutions such as the Voeikov Main Geophysical Observatory (GGO) and the State Hydrological Institute (GGI). Third, we provide an indication of the range of approaches to climate change amongst Soviet scientists encompassing the empirical approach of Mikhail Budyko and his colleagues through to the global change framework advanced by Kirill Kondratyev, and the natural climate change agenda of individuals such as Igor Maksimov (1910–1977).

Meteorology and climatology in the Soviet Union

Weather and climate have formed the object of concerted scientific action within Russia since the mid-nineteenth century (e.g. Oldfield, 2013). This interest embraced theoretical as well as applied aspects and reflected a number of competing concerns linked to the demands of a vast and growing empire. Furthermore, Russia's large land mass encompasses a range of latitudinal climates transitioning from the polar regions of its northern lands through to the hot desert environments along its southern borders, and this has invited speculation as to the origin and regularity of large climatic systems during the course of the last two centuries. More specifically, both meteorology and climatology played important roles during tsarist times due to the significance of weather prediction for enlarging and maintaining the empire, for improving agricultural activity and in the development of effective transportation networks and urban infrastructure.[3] Soviet scholars thus inherited a solid foundation of meteorological and climatological science from the pre-revolutionary period (e.g. Lydolph, 1971). Both branches of science gained influence during the course of the twentieth century impelled by the ambitions of the Soviet state. The need to understand the climate system in order to facilitate its management and regulation expanded with the demands of a rapidly growing economy. This was particularly apparent with respect to the country's agricultural sector. Concern over recurring drought episodes in Russia's European steppe region (the country's 'bread basket') extended back to the nineteenth century and formed a key motivation for the work of the soil scientist V.V. Dokuchaev (1846–1903) during the 1880s–1890s (see Moon, 2013). Primarily in response to the region's severe drought of 1891–1892, Dokuchaev devised a plan to transform the climate of the steppe with remedial action (Moon, 2013: 292–293; Johnson, 2015; Oldfield and Shaw, 2016), and elements of this work would re-emerge in the 1948 Great Stalin Plan for the Transformation of Nature, although in much modified form and hamstrung by the erroneous science of Trofim Lysenko (Brain, 2010: 673–675; Shaw, 2015). Nevertheless, the 'mastery of nature' thesis promulgated by the Stalin Plan was a common motif of the early- to mid-Soviet period and found expression throughout Soviet society. For example, the 'end goal of meteorology', as a Soviet textbook stated in 1960, was how to 'artificially change the weather and the climate in a direction preferable for humans [and to] reduce their dependency on the weather and on climatological conditions' (Skliarov, 1960: 8). At a general level, such thinking also underpinned research into weather modification, which expanded during this period in tandem with developments in the USA and elsewhere (e.g. see Gestwa and Belge, 2009; Fleming, 2010). This area of research activity also enabled both meteorologists and climatologists to demonstrate the relevance of their activities for key branches of the economy. Indeed, in an overview piece in the journal *Meteorologiia i Gidrologiia* (Meteorology and Hydrology) on the fortieth anniversary of the 1917 Revolution, Budyko suggested that the relatively rapid development of expertise in the general area of meteorology had been assisted greatly by the growing demands of the national economy, emphasising a highly intertwined relationship between the two areas (Budyko, 1957: 7). Similarly, M.S. Kulik's article in the same issue drew attention to the marked advancements evident within the specialist applied area of agro-climatology facilitated by the Soviet system (Kulik, 1957: 32). While the two articles are understandably celebratory in nature, they are nevertheless effective in highlighting positive trends evident within these branches of science post-1917.

The considerable expansion of the Soviet economy from the 1920s onwards and the evident importance of meteorology and climatology to such development prompted the establishment of a range of specialist institutions. Due to the limitations of space, the following section focusses on some of the most influential only.

Climatological and meteorological services of the Soviet Union

The Soviet Union's hydrometeorological services were unified during the course of the 1920s and 1930s in order to create a centralised body (Main Directorate for Hydrometeorological Services, henceforth Gidromet). This centralising process was primarily an effort to rationalise the country's network of air and water monitoring stations and associated data processing activities (Fedorov, 1967: 8–10). In time, much of the research related to weather, climate (including climate change) and the physical environment in general would be carried out under the auspices of this body. Its power increased with the growing need and desire of the Soviet government to monitor and control the environment evident from the 1970s (e.g. Kirillin, 1972: 2–3). To this end, the organisation was reformed in 1978 to become the State Committee for Hydrometeorology and Environmental Monitoring (henceforth Goskomgidromet) and thus gained more influence within the Soviet government. This shift also saw it made responsible for the USSR's national and international policies in related areas of environmental monitoring (e.g. see Sokolov et al., 2001: 150). The strengthening of the country's hydrometeorological services coincided with the growing importance and visibility of climate change science on both the domestic and international stage, and Goskomgidromet emerged to play a key role in the official handling of related issues during the course of the 1970s. Two former leaders of Goskomgidromet, E. K. Fedorov (who headed the Soviet delegation at the first World Climate Conference in 1979) and Yu. A. Izrael, were supporters of climate change research. At the same time, there is evidence to suggest that this support was dependent, particularly in the case of Izrael, on climate change research remaining within the remit of Goskomgidromet (Sokolov et al., 2001: 151). In order to do ensure this, Yuri Izrael founded the Moscow-based Laboratory for Monitoring the Environment and the Climate (LAM) in 1978, in which he gathered Moscow-based scientists from various relevant departments (Arkticheskaia entsiklopediia, 2017: 72). Three years later, this laboratory, together with 23 other scientific institutions belonging to either Goskomgidromet, the Academy of Sciences or the Ministry of Higher Education, were brought together within the remit of a State Scientific and Technical Programme (1981–1985) in order 'To develop methods for assessing possible climate changes and the impact of these changes on the national economy'. The programme was introduced under the auspices of the State Commission for Science and Technology.[4]

The Soviet Union's most important institution devoted to meteorological and climatological research was the Voeikov Main Geophysical Observatory (GGO) founded in St. Petersburg in 1849 during the reign of Nicolas I.[5] The initial establishment of the Observatory was given impetus via the encouragement of Alexander von Humboldt who had underlined the importance of general observational activities for a growing empire following his expeditionary work in Russia during the late 1820s (e.g. Humboldt, 2009: 266–288; Oldfield and Shaw, 2016: 37–40). The GGO's scientific activities expanded during the second half of the 1860s with the development of observation stations throughout the Russian empire, providing the basis for rudimentary weather forecasts from the early 1870s (Budyko, 1967: 3–4). Interest in the broader climate system developed strongly within the GGO and other institutions (such as the Russian Geographical Society) during the following decades underpinned by the work of individuals such as A.I. Voeikov (1842–1916) (e.g. see Voeikov, 1884; Kaminskii, 1916). During the early years of Soviet power, the Observatory's responsibilities encompassed a range of practical concerns (e.g. maintenance of the physical monitoring network) in addition to its longstanding academic work. However, the noted creation of a unified hydrometeorological service during the 1920s–1930s allowed the GGO to dedicate itself primarily to scientific pursuit (Budyko, 1967: 5). Dominant fields of research included dynamic, synoptic (large-scale) and theoretical meteorology in addition to both theoretical and applied areas of climatology (e.g. agro-climatology, bio-climatology).[6]

The GGO's profile was further enhanced during the Soviet period by the work of individuals such as M.I. Budyko, who rose rapidly within the Observatory post-1945 in order to lead a group of researchers concerned with quantitative climatology and, most influentially, the heat-water balance at the Earth's surface. This work deepened understanding of meteorological and hydrological processes and also proved influential in the West following the translation of Budyko's main work in the area.[7] Budyko was appointed director of the Main Geophysical Observatory in 1954, where he remained for the next 20 years. Together with his colleagues, Budyko would go on to advance a focussed research agenda concerning the study of global climate change. This focus evolved during the course of the 1960s–1970s out of developing interests in the causes of climate change, the sensitive character of the Arctic and, by extension, the global climate system and the potential to modify and transform the climate at a range of scales (e.g. see Gal'tsov, 1961).[8]

The growing interest of Budyko and his colleagues in climate change issues from the early 1970s was very much conversant with developments in the West as evidenced by their referencing of English-language work. More broadly, the science of Budyko was recognised as innovative in certain respects and this included his attempt to model the noted sensitive nature of the global climate system (Budyko, 1969) as well as his efforts to predict possible future climate change caused by CO_2 emissions (Budyko, 1972). Both areas of scholarship emerged around the same time as analogous work in the West.

A persistent engagement with Western scholarship was very much a characteristic of the GGO, and at the same time less evident in related research departments. The Observatory benefitted from its relative importance within the Soviet academy thus making its work more visible to foreign scientists. It was also assisted by Budyko's prominence within Western climate science circles, a standing that had developed strongly following his work on the global heat energy balance in the 1950s. His reputation was further enhanced in this regard by his co-chairing of the Climate Working Group (VIII) forming part of the 1972 *US-USSR Agreement on Cooperation in the Field of Environmental Protection*. Nevertheless, in spite of his international reputation and evident success in pursuing a research agenda around climate change, Budyko's presence at the Observatory was not appreciated by all. This was linked to a range of factors including his public statements about the inevitability of climate change and local Party resistance to his recruitment of a number of Jewish colleagues at the GGO.[9] Furthermore, in 1969, he was unsuccessful in his bid to become a full member of the Soviet Academy of Science, which made him vulnerable as a director of the GGO.[10] Under pressure, Budyko left the Observatory in 1975 in order to take up a position at the Leningrad State Hydrological Institute, taking with him the climate change research programme as well as a number of climatologists from the Observatory.

The State Hydrological Institute (GGI) can trace its history back to 1919, emerging as part of the re-visioning and expansion of the Academy of Sciences that took place during the early Soviet period. From 1930, the GGI was affiliated to the aforementioned centralised hydrometeorological service as part of a broader reconfiguration of activities in this area. Budyko's arrival at the institute precipitated the establishment of a Department on Climate Change and Atmospheric Water Circulation (1975), which Budyko would go on to head (Sokolov et al., 2001: 148). The remit of this department, which included climatologists such as K.Ya. Vinnikov, N.A. Efimova and G.V. Menzhulin, was to explore natural and anthropogenic factors behind climate change at both regional and global levels. It was also tasked with predicting the extent of anthropogenic climate change and determining the socio-environmental impact of such change (Shiklomanov, 2009: 24). In addition to staff and a developed research agenda, Budyko also brought an array of international connections to the newly formed department linked in particular to his activities with the aforementioned US-USSR Cooperation Agreement (e.g. MacCracken et al., 1990).

The activities of the Arctic and Antarctic Research Institute (AANII) also fell under the umbrella of the centralised hydrometeorological services post-1945.[11] In contrast to the research programme conducted in the GGO and the GGI, the work of this institution had an understandable regional focus as well as a strong emphasis on applied aspects such as the opening up of the North Sea Passage. From the early 1960s, following AANII's formal transferral to Gidromet, the AANII and the GGO began to consider each other as academic competitors in various fields, and this included climate change research (Sarukhanian, 2013: 111). For example, the institute was heavily involved in POLEX, one of the component projects of the Global Atmospheric Research Programme (GARP) (Sarukhanian 2013: 194). However, the general work of the AANII on climate change failed to gain the international attention enjoyed by Budyko and his group at the GGO or later at the GGI. The reason for this can be related, at least in part, to the noted regional emphasis of the AANII as well as its focus on the natural causes of climate change (e.g. solar activity), which ensured its contribution to the emerging international debate around anthropogenic climate change was muted. A further possible reason may be linked to the AANII's generally closed status due to the political sensitivities surrounding its work in the polar regions.

Understandings of climate change

The scientific discussions concerning climate change evident within the institutions outlined above and elsewhere within the Soviet academy included both anthropogenic and natural causation. Furthermore, different approaches to anticipating future climate changes were also evident, incorporating empirical as well as modelling techniques and thus mirroring analogous debates in the West. This section distils a number of general trends in order to highlight the varied nature of thinking in this area.

Anthropogenic climate change

It is tempting when analysing the debates around climate change to look for clear-cut positions and particularly with respect to the underlying causal factors of such change. Nevertheless, it is worth bearing in mind that the differing viewpoints typically revolved around the relative contribution of one factor or set of factors, rather than an outright rejection of a particular approach, a situation also to be found within the Western climate science fraternity. Budyko and his colleagues at the GGO (and subsequently the GGI), advanced anthropogenic causation as a key driver of climate warming from the 1970s onwards, and Budyko made one of the first attempts to predict the extent of global warming attributable to CO_2 in 1972 (Budyko, 1972). The prominence of Budyko in the work of GGO during the 1960s tends to disguise the existence of alternative viewpoints within the institution. Nevertheless, conflicting opinions were evident and these emerged strongly during the early to mid-1970s. For example, Budyko's argument of a warming trend was challenged by E.P. Borisenkov (1924–2005), former vice-director of the AANII, who would go on to replace Budyko as director of the GGO in 1975 and who sided with the global cooling discourse evident amongst some climate scientists on both sides of the ideological divide at that time (e.g. Fleming, 1998: 132). Nevertheless, in spite of such differences, in general there appears to have been a strong tendency to support anthropogenic climate change within the GGO.

Natural climate change

There was resistance to the emphasis on anthropogenic climate change within other parts of the academy, with significant efforts expended to advance natural causation as the main driver behind global climate shifts. The work of I.V. Maksimov, who lectured at the Leningrad Higher Naval Engineering School, is illustrative of this tendency and worthy of some additional comment. Maksimov was a geographer by training and his work was influential in furthering insight into both oceanic and atmospheric circulation. This focus mirrored similar interests evident amongst Western scientists, and his published work demonstrated a deep knowledge of such scholarship (Maksimov, 1970). As part of this, he placed emphasis on what he termed the Sun-Earth system, which amongst other things encompassed the sun's periodic changes in activity, an area that had long attracted the attention of climate scientists. He acknowledged the considerable complexities of this system and the difficulties of determining the precise nature of the link between the Earth's physical systems and climate (Maksimov, 1970: 12–13). In order to study the impact of the geophysical processes on climatic variations more effectively, he pushed for the establishment of a laboratory in 1963. This initiative was denied by Gidromet; however, a similar laboratory on the Sun-Earth relationship was subsequently founded in 1967 (Sarukhanian 2013: 106, 138). It should be noted that Maksimov did not necessarily ignore the potential for anthropogenic factors to influence climate, but his work with large-scale geophysical and cosmic forces ensured that such influence was understated at best in his work. More generally, while Maksimov's approach to climate change reflected a longstanding interest of Russian climatologists (and cognate scientists) in the physical nature of the climate regime (e.g. Berg, 1922; Shuleikin, 1941: 349–357), his work also challenged the geocentric (Earth-focussed) characteristic of much scholarship in this area in order to place the climate system firmly within a cosmic framework.

Climate and the biosphere

A further distinctive approach to the climate change issue is noticeable in the work of individuals such as the aforementioned K.Ya. Kondratyev, who was a lecturer in physics at the University of Leningrad. He would later be made rector of the university. More specifically, he emphasised the complex character of the Earth's physical systems and the need to try and understand climate within this broader context thus dovetailing with the earlier expansive work of individuals such as V.I. Vernadsky (1863–1945) concerning the biosphere. In this sense, his general approach overlapped with the debate around the biosphere concept pushed forward on the international stage by UNESCO and influential scientists such as G.E. Hutchinson from the late 1960s onwards, as well as the global system theorising of James Lovelock (Gaia hypothesis) (Oldfield and Shaw, 2013). Kondratyev was one of several influential Soviet scientists on the international stage and was held in high regard by the World Meteorological Organisation (WMO), being awarded the WMO prize in 1967 for his contributions to climate science with his works on radiative energy transfer in the atmosphere and meteorological satellite data. During the course of the 1980s, Kondratyev made a number of interventions in the climate change debate noting the influence of human activity on the climate system through such vectors as pollution and forest degradation. At the same time, he underlined the complexities of the climate system, the difficulties of predicting change, and the need for caution in reducing the climate debate to one centred solely on such indicators as changing CO_2 levels (Kondratyev, 1992: 325–329; see also Cracknell et al., 2008: 2446). As part of this, he reflected on the ability of the biosphere to absorb some of the excess CO_2 produced by human activity. His challenge to an overwhelming focus on CO_2 brought him into conflict with Budyko's general approach to this issue. Kondratyev was also an advocate of the use of natural analogues in order to inform understanding of contemporary climate change, and these included palaeoclimates in addition to the climate regimes of other planets (e.g. Kondratyev and Hunt, 1982).

Reconstructing past climates, modelling future climates

A general trend evident within Soviet climate discussions during the late Soviet period revolved around the use of past climates in order to assist future predictions of climate change at the regional level. In addition to Kondratyev's noted interest, this area of debate was also apparent in the work of Budyko and his colleagues and emerged as a dominant element of the Soviet Union's international response to anthropogenic climate change. More specifically, it was suggested that the use of a palaeoclimatic approach, in contrast to climate modelling, had the potential to provide finer detail with regard to future climate change at the regional level. This approach was advanced by the Soviet Union within the context of the *US-USSR Agreement on Cooperation in the Field of Environmental Protection* (MacCracken et al., 1990). At the same time, this emphasis on a palaeoclimatic approach was arguably encouraged by the limited computing power available in the Soviet Union compared to the West, a constraint which stymied Soviet developments in the area. In contrast, the US contingent was able to take advantage of its relatively advanced technical capabilities in order to push ahead an expansive computer-based statistical modelling approach. The evident split between the Soviet and US contingents should not, however, be overstated. Both parties advanced work with respect to past climates and modelling, and the concluding publication of the Soviet-US Working Group on climate underlined the synergistic nature of the two approaches (MacCracken et al., 1990). Nevertheless, the Soviet predilection for a palaeoclimatic approach surfaced as part of the preliminary work forming the basis of the first report of the IPCC that was published in 1990. Pushed by Yuri Izrael (chair of Working Group II – Impacts Assessment), the palaeoclimate approach was nevertheless side-lined in the final analysis due to concerns over the veracity of such an approach to predict future climate change and the dominance of a statistical modelling approach within Working Group I which dealt with the science of climate change (Houghton et al., 1990: xxv; Oldfield, 2018).

Soviet interest in the regionally differentiated nature of climate change boiled over into controversy during the late 1980s with the assertion by leading Soviet climatologists such as Budyko that increased levels of CO_2 had potential beneficial consequences since they would result in improved crop growth for certain regions of the globe (Miller and Pearce, 1989). Budyko had long flagged the potential positive aspects of climate change, and yet his statements in this regard had typically been couched within a measured scientific framework (e.g. Budyko, 1962b: 10). In contrast, his less guarded comments around this issue, evident from the late 1970s onwards, were understandably met with consternation by an international scientific and political community poised to push for international action in order to address anthropogenic climate change, and as such were roundly criticised.[12]

Concluding remarks

In this chapter, we have attempted to provide an overview of some of the main characteristics of the Soviet Union's engagement with the climate change debate post-1945. As part of this, we have stressed the longstanding nature of Russian work related to meteorology and climatology linked closely to the needs of a growing empire during the late tsarist and Soviet periods. The main part of the chapter focussed on a number of key Soviet institutions involved with climate issues and operating under the umbrella of the country's hydrometeorological services. The final part of the chapter explored the differing approaches to climate change and noted the presence of concerted activity with respect to understanding the processes behind both anthropogenic and natural climate change. Furthermore, the Soviet emphasis on the use of palaeoclimatic analogues to assist in the prediction of future climate change was noted. Importantly, the advancement of this approach was most evident at the international level and emerged clearly during the preparations for the first IPPC report in 1990. The high-profile nature of this report ensured that ongoing Soviet work in the related area of computer modelling was largely ignored. The Soviet Union's limited technical capabilities in this area, particularly in comparison with the United States, can be asserted as one reason for this. The broader context of the Cold War and the inherent competition between East and West characteristic of this period are others. The West's general reluctance to engage with Russian language materials provided a further barrier to the exchange of ideas (Oldfield, 2018: 47). Nevertheless, it would also seem to be the case that climate change research was pushed more robustly by the West, and most notably the United States, driven by a growing concern over the anticipated effects of climate change on the US economy.

The understanding and associated insights of Soviet scientists concerning climate change discussed in this chapter were nevertheless incorporated into their dialogue with relevant Soviet authorities. These bodies had interest in the possible consequence of anticipated climate change for Soviet society, although perhaps to a lesser extent than Western counterparts. In public discourse governmental officials such as E.K. Fedorov, former head of Gidromet, tended to acknowledge climate change due to human activity whilst simultaneously advancing ways of adapting to climate change via technical solutions (e.g. Fedorov, 1979). As such, there are no obvious calls for a cut in CO2 emissions during the 1970s and 1980s, with individuals such as Fedorov emphasising that a reduction in the use of fossil fuels would impede human development (Fedorov, 1979: 26). The emphasis on a technological fix is understandable within the context of a regime that had for decades promoted social progress with dominance over nature (e.g. see Josephson et al., 2013), and, in this light, climate change was an environmental problem to be understood and managed like any other. Climate change thus appears to have remained a somewhat ambivalent topic with respect to science-state interactions in the Soviet case. While it attracted a great deal of scientific attention, it received muted responses at the highest political levels. Furthermore, while Soviet scientists participated in global research projects dedicated to understanding and predicting climate change, the local consequences of a possible marked shift in climate regimes were not so obvious or clear-cut for the Soviet Union as they appeared to be in the West.

Notes

1 This chapter is based on early work linked to a UK Arts and Humanities Research Council project entitled: Soviet Climate Science and its Intellectual Legacies, AH/P004431/1. For more details, see: https://sovietclimatechange.wordpress.com
2 For an exception, see Sokolov et al. (2001); Oldfield (2016). For accounts of climate change research histories in an international context, see Fleming, J.R. and Janković, V. (eds.), (2011). Klima, *Osiris Second Series*, vol. 26, pp. 165–244.
3 For reflection on the connections between meteorology and empire, see: Mahony (2016); Williamson (2015).
4 Russian State Archive of Economy (RGAE) f. 9480, op. 12, d. 1513a.
5 It acquired its contemporary name during the course of the 1920s, transitioning from the Main Physical Observatory to the Voeikov Main Geophysical Observatory by 1929.
6 For an overview of the different fields, see Lydolph (1971).
7 The original Russian version was published in 1956, i.e. M.I. Budyko, 1956, *Teplovoi Balans Zemnoi Poverkhnosti*, Gidrometeoizdat, Leningrad. It was published in English in 1958 (translated by N.A. Stepanova).
8 See also TsGANTD f. 372, op. 11, d. 346. For examples of Budyko's developing work in this area, see Budyko, 1962a, 1967b.
9 For an example of his statements concerning the inevitability of climate change, see his interview with Spencer R. Weart, 25 March 1990, www.aip.org/history-programs/niels-bohr-library/oral-histories/31675
10 Interview with Jochen Kluge, (PhD student at the GGO 1966–1970), December 2017.
11 The Institute has its origins in the early Soviet period. It was reformed as the Arctic and Antarctic Research Institute in 1958.
12 See the interview with Budyko in *Der Spiegel* in which he speaks of a future 'greenhouse paradise', *Der Spiegel*, 1990 (1), pp. 143–147.

References

Arkticheskaia entsiklopediia. (2017) *Kultura, nauka, obrazovanie, religiia 2017*. Paulsen, Moscow.

Berg, L.S. (1922) *Klimat i Zhizn'*. Gosudarstvennoe izdatel'stvo, Moscow.

Brain, S. (2010) 'The Great Stalin plan for the transformation of nature', *Environmental History*, vol 15, no 4, pp. 670–700.

Budyko, M.I. (1957) 'Meteorologicheskie issledovaniia v SSSR', *Meteorologiia i Gidrologiia*, no 11, pp. 7–16.

Budyko, M.I. (1962a) 'Izmenenie klimata i puti ego preobrazovaniia', *Vestnik Akademii Nauk SSSR*, no 7, pp. 33–37.

Budyko, M.I. (1962b) 'Poliarnye l'dy i klimat', *Izvestiia Akademii Nauk SSSR*, no 6, pp. 3–10.

Budyko, M.I. (1967) 'Glavnaia geofizicheskaia observatoriia imeni A.N. Voeikova posle velikoi oktyabr'skoi sotsialisticheskoi revolutsii', in *Glavnaya Geofizicheskaia Observatoriia Imeni A.N. Voeikova za 50 Let Sovetskoi Vlasti*, Gidrometeorologicheskoe izdatel'stvo, Leningrad, pp. 3–10.

Budyko, M.I. (1967b) 'Izmeneniia klimata', *Meteorologiia i Gidrologiia*, no 11, pp. 18–27.

Budyko, M.I. (1969) 'The effect of solar radiation variations on the climate of the earth', *Tellus*, vol 21, no 5, pp. 611–619.

Budyko, M.I. (1972) *Vliianie Cheloveka na Klimat*. Gidrometeoizdat, Leningrad.

Collis, C. and Dodds, K. (2008) 'Assault on the unknown: the historical and political geographies of the international geophysical year (1957–8)', *Journal of Historical Geography*, vol 34, pp. 555–573.

Cracknell, A.P., Krapivin, V.F. and Varotsos, C.A. (2008) 'Preface to the special issue: the remote sensing heritage of Kirill Ya. Kondratyev', *International Journal of Remote Sensing*, vol 29, no 9, pp. 2445–2448.

Fedorov, E.K. (1967), 'Sovetskaia gidrometeorologicheskaia sluzhba k 50-letiiu velikoi oktyabr'skoi sotsialisticheskoi revolutsii', in E.K. Fedorov (ed.), *Meteorologiia i Gidrologiia za 50 Let Sovetskoi Vlasti. Sbornik Statei*, Gidrometeorologicheskoe izdatel'stvo, Leningrad, pp. 5–20.

Fedorov, E.K. (1979) 'Izmeneniia klimata i strategiia chelovechestva', *Chelovek i Stikhiia. Nauchnyi Populyarnyi Gidrometeorologicheskii Sbornik na 1980g*. Gidrometeorologicheskoe Izdatel'stvo, Leningrad, pp. 25–28.

Fleming, J.R. (1998) *Historical Perspectives on Climate Change*. Oxford University Press, Oxford.

Fleming, J. (2010) *Fixing the Sky: The Checkered History of Weather and Climate Control*. Columbia University Press, New York.

Gal'tsov, A.P. (1961) 'Soveshchanie po probleme preobrazovaniia klimata', *Izvestiia AN SSSR, Seriia Geograficheskaia*, no 5, pp. 128–133.

Gestwa, K. and Belge, B. (2009) 'Wetterkrieg und klimawandel. Meteorologie im Kalten Krieg', *Osteuropa*, vol 59, no 10, pp. 15–42.

Houghton, J.T., Jenkins, G.J. and Ephraums, J.J. (eds.). (1990) *Climate Change: The IPCC Scientific Assessment*. Cambridge University Press, Cambridge.

Humboldt, A. von, (2009) 'Speech given on 16. /28.11.1829 at the Tsarist academy of sciences', in E. Knobloch, I. Schwarz and C. Suckow (eds.), *Alexander von Humboldt, Briefe aus Russland 1829*. Akademie Verlag, Berlin, pp. 266–288.

Johnson, E. (2015) 'Demographics, inequality and entitlements in the Russian famine of 1891', *The Slavonic and East European Review*, vol 93, no 1, pp. 96–119.

Josephson, P., Dronin, N., Cherp, A., Mnatsakanian, R., Efremenko, D. and Larin, V. (2013) *An Environmental History of Russia*. Cambridge University Press, Cambridge.

Kaminskii, A.K. (1916) 'Klimatologicheskie trudy Aleksandra Ivanovicha Voeikova', in *Pamiati Aleksandra Ivanovicha Voeikova* (otdel'nyi ottisk iz toma LII (1916g.) *Izvestiia* Imperatorskogo Russkogo Geograficheskogo Obshchestva), Petrograd, pp. 8–13.

Kirillin, V.A. (1972) 'O merakh po dal'neishemu uluchsheniiu okhrany prirody i ratsional'nomy ispol'zovaniiu prirodnykh resursov', *Pravda*, September 20, pp. 2–3.

Kondratyev, K.Ya. (1992) *Global'nyi Klimat*. Nauka, Sankt-Peterburg.

Kondratyev, K.Ya. and Hunt, G.E. (1982) *Weather and Climate on Planets*. Pergamon Press, Oxford.

Kulik, M.S. (1957) 'Agrometeorologicheskaia sluzhba za 40 let', *Meteorologiia i Gidrologiia*, no 11, pp. 32–40.

Lydolph, P.E. (1971) 'Soviet work and writing in climatology', *Soviet Geography: Review & Translation*, vol 12, no 10, pp. 637–665.

MacCracken, M.C., Budyko, M.I., Hecht, A.D. and Izrael, Y.I. (1990) *Prospects for Future Climates: A Special US/USSR Report on Climate and Climate Change*. Lewis Publishers, Chelsea.

Mahony, M. (2016) 'For an empire of "all types of climate": meteorology as an imperial science', *Journal of Historical Geography*, vol 51, pp. 29–39.

Maksimov, I.V. (1970) *Geofizicheskie Sily i Vody Okeana*. Gidrometeorologicheskoe izdatel'stvo, Leningrad.

Miller, J. and Pearce, F. (1989) 'Soviet climatologist predicts greenhouse "paradise"', *New Scientist*, 1679 (26 August), p. 24.

Moon, D. (2013) *The Plough that Broke the Steppes: Agriculture and Environment on Russia's Grasslands, 1700–1914*. Oxford University Press, Oxford.

Oldfield, J.D. (2013) 'Climate modification and climate change debates among Soviet physical geographers, 1940s–1960s', *WIREs Climate Change*, vol 4, pp. 513–524.

Oldfield, J.D. (2016) 'Mikhail Budyko's (1920–2001) contributions to global climate science: from heat balances to climate change and global ecology', *WIREs Climate Change*, vol 7, pp. 682–692.

Oldfield, J.D. (2018) 'Imagining climates past, present and future: Soviet contributions to the science of anthropogenic climate change, 1953–1991', *Journal of Historical Geography*, vol 60, pp. 41–51.

Oldfield, J.D. and Shaw, D.J.B. (2013) 'V.I. Vernadskii and the development of biogeochemical understandings of the biosphere, c.1880s–1968', *BJHS*, vol 46, no 2, pp. 287–310.

Oldfield, J.D. and Shaw, D.J.B. (2016) *The Development of Russian Environmental Thought*. Routledge, London.

Sarukhanian, E. (2013) *Igor Maksimov. Ikh Imenami Nazvany Korabli Nauki*. Geo-Graf, St. Petersburg.

Shaw, D.J.B. (2015) 'Mastering nature through science: Soviet geographers and the Great Stalin Plan for the transformation of nature, 1948–53', *The Slavonic and East European Review*, vol 93, no 1, pp. 120–146.

Shiklomanov, I. (2009) *Gosudarstvennomy Gidrologicheskomy Institutu 90 Let, 1919–2009*. Gosudarstvennii Gidrologicheskii Institut, Leningrad.

Shuleikin, V.V. (1941) *Fizika Moria*. Izdatel'stvo akademii nauk soiuza SSR. Leningrad/Moscow.

Skliarov, V.M. (1960) *Meteorologiia i Meteorologicheskie Nabliudeniia (obshchedostupnyi kurs)*. Gidrometeoizdat, Leningrad.

Sokolov, V., Jäger, J., Piarev, V., Nikitina, E., Ginzburg, Alexandre Goncharova, Elana, Cavender-Bares, Jeannine, and Parson, Edward A. (2001) 'Turning points: the management of global environmental risks in the former Soviet Union', in Social Learning Group (eds.) *Learning to Manage Global Environmental Risks, Volume 1: A Comparative History of Social Responses to Climate Change, Ozone Depletion, and Acid Rain*. MIT Press, Cambridge, pp. 139–165.

Voeikov, A.I. (1884) *Klimaty Zemnogo Shara v Osobennosti Rossii*. Kartograficheskoe zavedenie A. Il'ina, St. Petersburg.

Weart, S.R. (2003) *The Discovery of Global Warming*. Harvard University Press, Cambridge.

Williamson, F. (2015) 'Weathering the empire: meteorological research in the early British straits settlements', *BJHS*, vol 48, no 3, pp. 475–492.

3 Environmental change and the Soviet media before 1986

Dissident and officially sanctioned voices

Anna Mazanik

Introduction

Historians have documented the great interest of the Soviet public in environmental problems in the late 1980s. During the time of perestroika, environmental topics often appeared intertwined with anti-Soviet, nationalist or separatist discourses (Dowson, 1996). But where did this interest of the late 1980s come from? Was it triggered by the Chernobyl catastrophe? Or was it a continuation of the earlier debate, and, if so, how much of it had taken place underground, pushed out from the official press by censorship, before glasnost opened an arena for discussion? To what extent was Soviet public interested in global and local environmental issues before the 1980s?

Recently, the 'ecocide' narratives that pervaded much of the academic literature in Soviet environmental history have come under revision. A number of scholars have shown that the ruthless 'conquest of nature' and environmental catastrophes were only one side of the story. They offered a more nuanced interpretation by highlighting the areas of Soviet innovation and positive engagement with the environmental cause. For example, Douglas Weiner revealed how much open discussion and social mobilisation was taking place in the Soviet Union around environmental issues (Weiner, 1999). Nicholas Breyfogle argued that Baikal environmental movement in the 1950s gained momentum before any similar environmental protest started in the West (Breyfogle, 2015). Denis Shaw and Jonathan Oldfield investigated the role that Soviet scientists played in setting up the climate change agenda (Show and Oldfield, 2016). Most of the existing scholarship, however, focused on the activist groups and scientific debates. But how much would an average Soviet reader, that is someone not specifically interested in natural sciences or nature protection, be exposed to these debates? What was communicated to her by the media, through which channels and from which perspective?

This chapter studies the communication of environmental issues in the Soviet media in the period from the late 1960s to the early 1980s. In the first part, I will look at how the environmental problematique appeared in the official Soviet press, which issues received specific coverage and from which angle. The second part will focus specifically on the question of climate change and the communication of climate science as it happened on the pages of Soviet newspapers. However, the official sources did not capture the entire media landscape of late socialism. At least from the 1960s, the Soviet public had developed an alternative channel of communicating the uncensored information known as samizdat. Samizdat was successfully used to discuss topics banned from the official Soviet mainstream and therefore could have potentially served as a free forum for alternative environmental narratives. The last part of the chapter studies how important environmental problems were for the Soviet samizdat media and how different the uncensored narratives were from the discussion in the state-controlled forums.

This chapter is based on the large Soviet media and samizdat collections of the Radio Free Europe/Radio Liberty Research Institute held at the Blinken Open Society Archives (OSA). An American radio station, Radio Liberty started broadcasting to the Soviet Union in 1953 with the support of the CIA. The purpose of the RFE/RL Research Institute was to collect, analyse and verify all available sources of information about life behind the Iron Curtain that could be used to prepare the radio broadcasts. The Institute acquired major central Soviet newspapers and journals and some regional ones from the late 1950s to the early 1990s, which were then clipped and arranged according to the subjects they covered. In addition, since the late 1960s, the Institute started to gather all nonfiction samizdat materials smuggled to the West, which resulted in one of the world's largest samizdat collections. Although these press and samizdat collections are clearly mediated through the gaze of the RFE/RL staff and do not contain all the documents produced in the USSR, they are nevertheless highly representative and allow a researcher to study how specific problems were covered in various print media over long periods of time.[1]

Environmental problems and the official print media

Soviet media is often imagined as strictly censored and controlled, a mere instrument of the state propaganda projecting the monolithic vision of reality that the party officials wanted their society to have. A number of scholars have challenged this pessimistic view, showing that the party control of Soviet journalism still left space for critical investigation, dialogue and strong professional identities among some groups of journalists whose priorities deviated from those of the party (Wolfe, 2005; French, 2014; Huxtable, 2018). My own research led me to similar conclusions. No doubt, there was no freedom of speech in the Soviet Union, but the media communication clearly was not monolithic and a relatively free discussion on certain topics, with confronting and critical opinions, was still possible. Douglas Weiner (1999) has famously described Soviet environmentalism as a 'little corner of freedom', and that freedom was enjoyed not only within activist organisations and scientific institutions, but also to some degree in the press.

First of all, environmental policy in the Soviet Union, scattered among several governmental agencies, was generally a disputed domain and this affected its representation in the press. Regions of the Soviet Union competed for resources and could have very different stakes in each specific environmental question. All parties in the conflicts used media, both central and local, to justify their failures, shift responsibility or promote their cause. In addition to being the mouthpiece of conflicting governmental agencies, already in the late 1950s, Soviet print media emerged as a powerful tool of the nascent environmental lobby. In the Khrushchev era, the coverage of deteriorating environmental conditions went beyond the few scientific journals to a much larger audience. In the late 1950s and early 1960s, newspapers and journals such as *Literaturnaia Gazeta*, *Komsomol'skaia Pravda*, *Novyi Mir* and *Nash Sovremennik* commissioned their journalists to prepare investigative reports on these questions or opened their pages to concerned scientists (Kelley, 1975).

To give an example, in 1958, *Literaturnaia Gazeta* printed a letter signed by many prominent Soviet scientists, engineers and writers that marked the beginning of a campaign in defence of Baikal. The signatories fiercely opposed the proposal to organise explosion at the point where Baikal flows into the Angara River and to use the produced water rush to generate electricity. The letter described such catastrophic consequences of this measure that the plan was dropped (Breyfogle, 2015). This success provoked a wave of other publications in Soviet periodicals, most notably, *Literaturnaia Gazeta* and *Komsomol'skaia Pravda*, about other dangers faced by the lake, such as industrial development, untreated wastes, over-fishing, poaching and wood-logging (OSA 300–80–1:719/3). In 1962–1963, *Literaturnaia Gazeta* published a series of articles by writer (and former hydrologist) Sergei Zalygin against the construction of the lower Ob' hydropower station that would entail flooding an area of more than 100,000 sq. kilometres, a plan that too was eventually abandoned. Another gargantuan Soviet scheme, the river diversion project, was criticised in the media in the early 1960s. In 1963 (no.3), the literary journal *Nash Sovremennik* (the future stronghold of the 'village prose') published a polemical article by

Iu. Telitsyn 'Careful, Nature!', which warned against hasty decisions that could damage the economy of entire regions. Rerouting waters from the Pechora and Vychegda rivers to the Caspian, he wrote, would bury large woodlands and agricultural areas under water and disturb the climate of the Arctic coast (OSA 300–80–1:649/1).

Interestingly, in the time of Khrushchev, it was primarily the literary and youth periodicals that raised public awareness of environmental risks posed by the major construction projects. These media were essentially elit-ist, targeting intelligentsia and youth audiences, the social groups which were perhaps the closest to the forming circles of environmental activ-ists. In the Brezhnev era, the media coverage of environmental problems expanded dramatically. This can be seen in the frequency and the size of the articles published, their thematic diversity and the number of media outlets that were now working with the topic. In addition to the literary periodicals, from the turn of the 1970s, anthropogenic impact on the envi-ronment was regularly discussed on the pages of the major Soviet dailies, including *Pravda* and *Izvestiia*, periodicals of the trade unions (*Trud*), the party (*Sovetskaia Rossiia*), the army (*Krasnaia Zvezda*) or regional news-papers, which were reaching a much broader audience. Some newspapers established special environmental rubrics, for example, 'Man and Nature' in *Pravda* and *Sovetskaia Rossiia* or 'Nature and People' in *Izvestiia* and its weekly addition *Nedelia*. Several factors must have contributed to such a surge of interest.

First, in this period, nature protection and pollution abatement took a prominent place in the state rhetoric (Josephson et al., 2013). The country made an effort to establish a modern legal system to mitigate the costs of environmental degradation. New all-union water legislation was adopted in 1970 and the RSFSR Water Code two years later. In 1972, the USSR Supreme Soviet approved a decree 'On the Enforcement of Nature Protec-tion and the Improvement of Usage of Natural Resources' and a number of similar acts were promulgated by other party and state agencies. That legis-lation, although rarely implemented, revealed the growing concern for the environmental issues among state officials – or at least their perceived need to show such concern – and gave a 'green light' for the media to cover envi-ronmental topics. Soviet newspapers eagerly reported on the newly adopted decrees, the discussions that preceded their promulgation, and recommen-dations on the further development of environmental legislation. For exam-ple, an article by lawyers G. Aksenenok and N. Syroiedov (*Pravda*, 1973), 'Law Protects Nature', called for the need to coordinate environmental law on the all-union level and criticised the practice of keeping the functions of nature exploitation and protection within the same ministry (OSA 300–80–1:720/1). In another article ('Limits for Emissions', *Pravda*, 7 June 1975), a group of prominent scientists demanded a law that would set the maximum permitted levels of dangerous industrial emissions (OSA 300–80–1:720/1).

This was also the time when the environment had become an issue of global importance. The Soviet Union was involved in the growing international cooperation in the field, helped by the beginning of the détente in the Cold War politics. At the 1972 summit meeting in Moscow, the Soviet Union and the US signed an agreement on environmental cooperation which oversaw extensive joint programmes under 11 major areas, ranging from air, water and soil pollution control to earthquake prediction and the study of climate (OSA 300–80–1:720/2). Soviet press reacted to major international nature protection events (for example, the celebration of the World Environment Day in 1973) and closely followed the development of the Soviet-American cooperation. Usually, the articles on this topic argued that the cooperation across the Iron Curtain was the only way to save the planet from further environmental degradation. Although acknowledging the differences between the US and the Soviet Union, they promoted the image of an equal, positive and productive partnership and treated American expertise and experience with respect and appreciation, sometimes even inviting the commentary of American scientists.

In the 1970s, the articles on global environmental problems, especially pollution, resource depletion and threats to biodiversity, emerged as a somewhat separate genre. Relatively long, with titles such as 'Is Crisis Unavoidable?' (*Komsomol'skaia Pravda*, 29 March 1975) or 'Biosphere Is Our Home' (*Sovetskaia Rossiia*, 6 June 1973), they usually had a format of an interview with famous scientists or were directly written by them. Giving the floor to scientists ensured that such articles had a reserved, semi-academic style, presented balanced and well-argued opinions, avoided bold statements and spoke with caution about what can be known and done. They could sharply contrast with other texts in the same newspapers, with their overly optimistic tone and exorbitant praise for Soviet achievements. Some print media even staged roundtable discussions and invited several scientists from various disciplines, giving them the opportunity to debate global ecological challenges from different perspectives, for example, a discussion panel 'Global Ecology: A New Science?' in *Literaturnaia Gazeta*, 24 January 1973 with 14 scientists, including cybernetist Aksel Berg, physicist Pyotr Kapitsa, geophysicist Evgenii Fedorov, climatologist Mikhail Budyko and geneticist Iurii Rychkov (OSA 300–80–1:720/1).

However, a new interest in global problems, which became evident in the Soviet media at the turn of the 1970s, did not mean that the local environmental causes were forgotten. The major concerns of the Khrushchev period – the protection of Baikal and the river diversion project – continued to receive media attention and were now covered also in the more 'democratic' newspapers, such as *Pravda, Izvestiia* or *Trud*. In addition, several new environmental campaigns were supported by the press, for example, the protection of Lake Sevan, of the Desna River or of the taiga threatened by the construction of the BAM railway (OSA 300–80–1:720/1). Local stories were often used to address more general environmental problems that permeated Soviet practices of economic management and everyday life. Throughout the 1970s and early 1980s, Soviet newspapers critically discussed the lack of strategic planning for sustainable resource usage and protection, water pollution by agriculture and industry, inadequate financial and administrative support for pollution abatement, poaching and destructive mass tourism in natural reserves, inefficiency of existing regulations and fines or public health problems connected to pollution. Although scientific commentary was sometimes used to write about such topics, more common formats were either investigative reports by newspapers' own journalists or the letters of 'concerned citizens', usually local teachers, engineers, lawyers or representatives of the 'people's control' organisations. Newspapers often made responsible authorities comment on those letters, which sometimes led to local investigations or even to the punishment of violators (OSA 300–80–1:719/5; 300–80–1:720/1–4; 300–80–1:721/1–3).

Certainly, environmental reporting in the Soviet press under Brezhnev had its limitations. Anything related to security matters was classified; information about environmental disasters and the scale of pollution was kept secret. Even with the less 'dangerous' topics, there were clear instances of direct censorship. In 1970, the party imposed a temporary ban on the discussion of environmental problems, leading *Izvestiia* to cancel a proposed series on the pollution of the Volga (Kelley, 1975). In 1973, information about the river diversion project was restricted (Josephson et al., 2013). Articles on global ecological problems in Soviet newspapers were much more likely to bring examples of environmental negligence from the capitalist block and never provided any sufficient statistical data that would allow Soviet readers to see how the USSR performed in comparison to the rest of the world.

Yet, from the turn of the 1970s, any regular reader of the Soviet press received sufficient information to see that environmental management in the USSR was highly problematic, even if the picture presented was incomplete. It is difficult to judge how important the role of the media was for specific environmental campaigns, but it is evident that the major *causes célèbres* of the Soviet environmental movement did receive coverage, sometimes rather significant, in the official press. And even if such publications did not bring effective policy changes, they did raise awareness about global, national and local environmental problems.

Communicating climate change in the Soviet press

The interest of the Soviet media in the questions of global weather and climate started in the 1960s, triggered by the worldwide development of satellite meteorology and the related intensification of the international cooperation in that field with the involvement of the USSR, for example, the creation of the World Weather Watch in 1963 and of the Global Atmospheric Research Program in 1967. The launching of the Soviet weather sputniks and the spacial meteorological system in 1966–1967 was proudly celebrated in the Soviet press as a major achievement and the USSR's service to the global community (OSA 300–80–1:673/5).

Extraordinary weather events have always appeared in the news, both in the Soviet Union and elsewhere, but already from the early 1960s, one could find occasional reports on the possibility of a more structural climatic change responsible for the unusual conditions. In 1961 (8 December), *Trud*, the main newspaper of the Soviet trade unions, gave the following explanation to the abnormally mild winter (OSA 300–80–1:673/5):

> It is possible that recently a new reason for a steady warming appeared. The development of cities and rapid industrial growth require high consumption of fuel. When this fuel is burnt, huge quantities of carbon dioxide are released into the atmosphere. And although the gas concentration in the atmosphere is no more than several hundredths of a per cent, it plays a very important role – it keeps the Earth from emitting excessive warmth to space through radiation.

Yet, it was not until the turn of the 1970s that climate change attracted the sustained attention of the mass Soviet press, the same time when it started to emerge as a public concern worldwide. In the Soviet media, the interest could have also been sparked by the extremely hot and dry summer of 1972 in Central Russia, accompanied by severe drought and thousands of forest and peat bog fires, which for weeks kept the Moscow region in a thick smog. The event was highly damaging for Soviet economy, as it destroyed a huge part of the country's food crops, forcing it to buy up US wheat reserves; dozens of settlements and large woodlands were lost to fires. The heat wave stayed in the public memory as one of the most severe weather events of the century and thus could have stimulated journalistic interest in long-term weather and climate development.

In 1972 (25 June), *Pravda* published a long article entitled 'The Climate of the Future'. The author of the article was an internationally famous Soviet climatologist Mikhail Budyko (discussed in Chapter 2). Budyko's article was one of the early texts in the mass Soviet press that focused specifically on climate change. In this article, Budyko mentioned the historic human influence on the climate regimes through deforestation and irrigation but stressed that it is the energy production, the burning of fossil fuels and the rise of CO_2 in the atmosphere that had the major impact on global climate in the twentieth century. He warned that it is not only the possibility of climate change but also its rapidity that pose the major challenge: 'With today's speed of technological development, it will start influencing economic activity not later than in 20–30 years, and in 50–80 years will dramatically change it in many countries'. Budyko also explicitly stated that such a quick change of climate would create enormous difficulties for large regions of the globe and pose complex international problems. To avoid these negative consequences, he warned, humankind would need to change the character of its economic activity well in advance (OSA 300–80–1:674/1).

Throughout the 1970s and early 1980s, it was almost exclusively through the mouth of scientists that climate change was discussed in the Soviet press. The questions were formulated by the journalists or taken from the letters of the readers – who, as the editors regularly claimed, were very interested in the issue – but it was scientists who provided the answers. Apart from Budyko, who repeatedly appeared in newspapers, the list of other invited scientists included, for example, the Director of the Soviet Hydrometeorological Center Mikhail Petrosiants ('Is the Climate of the Planet Changing?', *Izvestiia*, 12 March 1974), paleoclimatologist and the Head of the Paleogeography Department in the USSR Academy of Sciences Andrei Velichko ('What Will Climate Be Like?', *Izvestiia*, 21 August 1975), Deputy Director of the Hydrometeorological Center V.N. Parshin ('Is the Atlantic Guilty?', *Pravda*, 19 October 1976), atmospheric physicist Kirill Kondratyev ('Weather Kitchen', *Sovetskaia Rossiia*, 1 April 1979 and 'Climate of the Earth', *Izvestiia*, 4 May 1980), geophysicist and the head of the Soviet delegation to the World Climate Conference Evgenii Fedorov ('Climate and Humans', *Izvestiia*, 23 March 1979) (OSA 300–80–1:674/1–2).

Although most of the articles on climate change warned of the likeliness of global warming due to the increase of the CO_2 concentrations in the atmosphere, the global cooling scenario was also mentioned. A wave of interest in this hypothesis was connected to the 1974 CIA report on the relations between climatology and intelligence, which through the angle of the Cold War politics speculated how global cooling would affect the economic potential and the distribution of resources between the two world powers. The report brought the issue to the attention of the media. If global warming was sometimes believed to be potentially beneficial to the Soviet Union and more dangerous to the US, global cooling was expected to have the opposite effect (see, for example, 'CIA's Report on World Climate Predicts a Catastrophic Change', *IHT*, 3 May 1976, or 'Colder Era Is Bad

News for Soviet Union', *IHT*, 5 October 1976, OSA 300–80–1:674/1). The idea of global cooling was highly controversial and had many opponents among climatologists. Disproving its conclusions could have worked well not only to promote Soviet science and its views of the problem, but also to serve the Soviet cause more generally through discrediting the CIA. This risk was soon recognised by the Western propaganda institutions when the Radio Liberty staff was advised not to discuss global cooling in their broadcasts (OSA 300–80–1:674/1). Soviet press commented on the CIA report and refuted its conclusions as not having enough scientific evidence (see the article by the Vice-President of the World Meteorological Service Iurii Izrael 'Is Climate Getting Worse?', *Literaturnaia Gazeta*, 12 January 1977, OSA 300–80–1:674/1).

Unlike the Western media, Soviet journalists reported on climate change only through a scientific lens and avoided writing any original investigation or opinion pieces on the topic without hiding behind the direct speech of scientists. The rare exceptions to this canon of 'scientific commentary' were translations from the Western press. For example, in 1974 (11 October), *Trud* republished an article from West German magazine *Der Spiegel*, summarising major scientific theories on climate change. In 1977 (14 December), *Literaturnaia Gazeta* translated another piece from *Der Spiegel*, an alarmist article on the dangerous consequences of the greenhouse effect, and invited three Soviet climatologists and meteorologists, Mikhail Budyko, Feofan Davitaia and Iurii Chirkov, to comment on it. All commentators agreed on the likelihood of the global warming scenario and the dramatic impact it would have. Budyko and Davitaia, however, distanced themselves from the pessimistic conclusions of *Der Spiegel*, referring to the possible technological solutions that could help humanity prevent or mitigate the negative effects of excessive warming. Chirkov, on the other hand, did not accuse *Der Spiegel* of undue alarmism and stated that the consequences of the 'global greenhouse effect will be, without doubt, harmful for the traditional forms of agricultural organization'. (OSA 300–80–1:674/1).

Reporting on climate change only within the framework of science determined the image of the problem on the pages of Soviet newspapers. First, the communication of the problem was almost devoid of any sensationalism and apocalyptic visions. Monocausal explanations, bold statements and definite predictions were avoided. Almost every scientist invited to comment on the issue mentioned the complexity of the problem, the limitations of existing data and discussed various factors that could influence the climate. Although anthropogenic global warming was a favoured hypothesis, it always appeared as one of many possible explanations. The rhetoric varied from moderate alarmism to moderate optimism, depending on the individual position and expertise of scientists.

Secondly, most authors showed a strong commitment to the values of peace and international cooperation. Of all the articles on climate change that I managed to identify, not even one used the issue as a pretext to offer a critique of the West or of capitalism more generally. Even when there were specific reasons for easy criticism and polemics – for example, when commenting on the CIA report on global cooling – the rhetoric remained moderate and academic in style. Soviet scientists always wrote about their Western colleagues with respect and appreciation and almost never invoked that they belonged to different ideological camps. In their presentation, climate change was a global issue, which was studied and should be dealt with by the global community together. It also meant that the idea of differentiated responsibility for global warming had not yet appeared. In respect to climate change, the world was not divided into the West and the East, the North and the South. It was 'humankind' as a whole that was responsible for anthropogenic climate change and that should work together to prevent its consequences.

Finally, the focus on the global scale of climate change left no space for talking about individual responsibility. The issue appeared too enormous to imagine that individual choices and local practices could influence it. The state seemed to be the lowest level on which the climate effects can be determined – and the state that the authors of the articles had in mind was clearly a big and powerful one, capable of implementing gargantuan schemes of nature transformation. Although scientists addressed their message to the international community, they often implicitly spoke to the Soviet government too, warning it against hasty steps that could influence the regional and global climate, such as the river diversion project, and arguing for continuous support for international cooperation.

Environmental narratives in the underground media

The information landscape of the Soviet Union was not reduced to the official media only. Critically minded Soviet citizens of the Brezhnev era, however small their circles might have been, bravely used other sources of information such as foreign broadcasting and underground samizdat publications.

Although foreign radio stations helped inform Soviet citizens about the Chernobyl catastrophe, their role in raising awareness of environmental issues in earlier periods was less important. In the *Radio Liberty* broadcasts, the topics of nature protection did not appear until the early 1970s, that is, only after they started to be actively discussed in the official Soviet press. Environmental reporting throughout the early 1970s maintained a rather moderate tone and the overall impression their programmes conveyed did not significantly differ from the rhetoric of the critical Soviet publications on environmental issues. The internal reports of *Radio Liberty* in the early 1970s, in fact, held a rather favourable view of the Soviet responses to environmental challenges; they concluded that, despite many ecological problems in the USSR, the situation there was not worse than in the West and suggested that some practices (for example, the extensive usage of electric public transport) should even be adopted in the US – although this was of course not reported in the Radio's broadcasts (OSA 300–80–1:719/5). It was only towards the turn of the 1980s that the *Radio Liberty* discussion of the environmental problems in the Soviet Union turned more negative and focused on the narratives of 'ecological crisis' (OSA 297–0–1).

As for the samizdat publications, a researcher may be somewhat disappointed to find the scene of environmental activism in the times of stagnation relatively empty, especially in comparison with the diverse green movements of the perestroika. We know of dozens of unofficial ecological groups of the late 1980s, but the RFE/RL archives preserved no memory of any dissident group of the 1960s and 1970s whose primary concern was the environment. There were a number of human rights groups, nationalist, religious, literary, pacifist and disability groups, but if any organised dissident environmental group existed at the time, it failed to make connections with the West or other underground movements in the Soviet Union and was not noticed by the vigilant RFE/RL observers. This does not mean that environmental issues were not covered in the underground publications but this topic appears far less important in the Soviet samizdat than religion, freedom of emigration, abuse of psychiatry, nationality rights and democratisation. The Samizdat Archives of the RFE/RL Research Institute (OSA 300–85) contain about 6,000 Samizdat documents produced between 1965 and 1985, and only several dozen of them, or about 1%, discuss environmental questions in some form or another. However, not all samizdat publications had the same influence and circulation and, in fact, environmental topics did appear in some of the most important documents of the time.

One such document was Andrei Sakharov's 1968 essay 'Reflections on Progress, Coexistence and Intellectual Freedom'. The essay had a wide circulation in samizdat and, after another Soviet dissident, Andrei Amalrik, gave a copy of the essay to a foreign correspondent, it was published in the West, and within one year, 18 million copies were distributed across the world (Bergman, 2009). In the essay, Sakharov focuses on three major global challenges: the threat of nuclear war, the threat of hunger and the question of 'geo-hygiene'. Under the last point, he includes a variety of environmental problems, such as air and water pollution, pesticides, uncontrolled use of antibiotics in livestock raising the risk of drug-resistant bacteria, radioactive wastes, soil erosion and salination, extensive wood-logging, extinction of many species and climate change, or, as he calls it, 'overheating of the Earth' because of the CO_2 emissions (OSA 300–85–9:5/45).

For Sakharov, these were global problems that could only be solved if nations and governments act together – a conviction that was clearly shared by many Soviet scientists. The primary solution for him was the avoidance of a nuclear war, the recognition of the commonality of human interest and convergence of the capitalist and socialist political systems. Sakharov's text was written before the discussion of global ecological challenges became common in the Soviet newspapers. Although no similar calls for convergence of socialism and capitalism could have appeared in the official media, the idea that cooperation across the Iron Curtain was the only way to solve the world's environmental problems would be repeatedly brought up by the Soviet newspapers in the following years and also receive some practical realisation.

Sakharov saw the Earth primarily as a resource that should be used for human needs. He maintained that science and human innovation was the only way to overcome hunger and poverty and to deal with environmental challenges. This prometheanism and the belief in progress and human reason also relate Sakharov to other Soviet scientists communicating through official channels – which is perhaps not surprising considering that he was a product of the same milieu. When talking about global environment and scientific progress, Sakharov, long before it became commonplace, saw the world in terms of North and South, claiming that it was the duty of the technologically advanced North to help the South (Bergman, 2009). Although this differed from the more unitary vision of humankind vis-a-vis global environmental challenges that would be presented by scientists in the official Soviet press, it is quite remarkable that all interpretations implied that the Iron Curtain was not a particularly important line in the context of ecological impact and scientific innovation.

Sakharov developed his points further in another text 'The World in 50 years' (published in tam- and samizdat), arguing that there is no option of 'turning back' and further technological progress is the only way to overcome ecological challenges and avoid resource depletion. He envisaged intensification and new forms of agriculture, including the conversion of the surface of the oceans for agricultural purposes, an increasing role for genetic modification and fertilisers, artificial production of animal proteins, the greater use of coal and nuclear energy (OSA 300–85–9:47/20). At the turn of the 1970s, Sakharov's points and, in particular, his diagnosis of the global environmental problems, were briefly repeated in some other samizdat documents, for example, 'The program of the Democratic Movement of the Soviet Union' (1969, OSA 300–85–9:8/30). However, unlike Sakharov's other concerns – such as the freedom of information and human rights – the environmental problematique did not achieve much resonance in the so-called 'democratic' Soviet samizdat and in the 1970s practically disappeared from its agenda.

In the early 1970s, the interest in environmental problems was more visible in the so-called 'conservative' samizdat. One of the important samizdat media here was the journal *Veche* published between 1971 and 1974 (all together 10 issues) by historian Vladimir Osipov. He imagined it as an 'apolitical' journal, which would cover such 'apolitical' topics as Russian history, the Orthodox Church, cultural heritage and environmental protection. To show his intention to stay within the allowed political boundaries, Osipov even put his name and address on the cover of the journal, which did not save him from arrest and eight years in the camps. The journal featured occasional texts on local ecological problems in the Soviet Union, which, in fact, were quite similar in content and style to the articles in official Soviet newspapers. The most elaborate analysis of environmental problems appeared in the third issue of *Veche* (1971), connecting global ecology and Russian patriotism. Already in the editorial, Osipov warned his readers about the dangers of constant industrial growth, pollution, overuse of pesticides and cars, ending it with the following words: 'We call all Russian patriots to think about the coming catastrophe. One sixth of the dying planet is our motherland. We ask: is there time to save Russia?' This question received further elaboration in an anonymous article in the same issue (probably, also written by Osipov), 'The House We Are Building'. In this rather lengthy and detailed text, the author draws a gloomy picture of global and Soviet environmental problems, using a variety of sources, from scientific papers, to the materials of the WHO and UNESCO to Rachel Carson's 'Silent Spring' and Jean Dorst's 'Before Nature Dies'. Seeing the Soviet experience as a part of this international development, the author warns his readers against competition with the US and fascination with consumerist culture, suggesting instead that humankind should slow down and limit its growth (OSA 300–85–9:28/26).

Probably, the most important document on environmental issues from the 'conservative' samizdat was Alexander Solzhenitsyn's 'Letter to the Leaders of the Soviet Union' (1973), the text that was in dialogue with Sakharov's 'Reflections on Progress' and was often analysed together with it. In this letter, Solzhenitsyn warns of the imminent national catastrophe that could come either from a war with China or from an environmental collapse, the 'all-encompassing dead-end of the Western civilisation which Russia chose to be a part of'. In Solzhenitsyn's interpretation, Russian thinkers, like their fellow colleagues in the West, were trapped in the Enlightenment illusion that constant progress is a possibility and a necessary goal, without realising that the resources of the planet are finite, and now 'the greedy civilization of eternal progress reached its end', as the report of the Club of Rome has demonstrated. Solzhenitsyn claims that a centrally planned economy actually could have provided a good chance to prevent the destruction of nature, but Russia did not use this chance, depleting its resources and turning large territories into an industrial wasteland. He gives very little credit to science in general and Russian science in particular. But, he says, there is still a way out because, in the current environmental situation, land is the greatest wealth: 'This is our Russian hope for salvation: on our wide north-eastern territories, which we have not yet spoiled, we can build not a mad and devouring civilization of progress, but a sustainable economy, and settle people there according to its principles' (ibid). Although many contemporary commentators viewed Solzhenitsyn's principles of that new social existence as very conservative and backward, when seen from the twenty-first century, some of them might actually appear quite progressive. He recommended closing most polluting factories, especially in the military sector, limiting urbanisation and returning to smaller green towns with low-rise housing and abundant gardens where only electric transport is allowed, or as he puts it, 'towns not for cars but for people, horses, dogs and trams'. Soviet leaders, Solzhenitsyn concluded, should initiate these changes with their own hands, without waiting for the revolution, because they had enough resources to organise this gradual transition and to ensure Russia's national revival (OSA 300–85–9:42/19).

The debate between Sakharov and Solzhenitsyn on global challenges and scientific progress was the high point in attention to environmental issues among Soviet dissidents. It occurred during the short 'golden era' of Soviet samizdat, the time of its widest dissemination, broadest thematic coverage and high hopes for change among its creators. In the mid-1970s, the dissident movement experienced a certain crisis, connected to the fierce repression, imprisonment or exile of the main activists and the realisation that all their efforts were to no avail. As a result, in the late 1970s–early 1980s, the dissident movement concentrated primarily on reporting the human rights abuse. Although the 1975 Helsinki Agreements, the main reference document for many dissidents, had an entire section on environmental protection, Soviet environmental and human rights movements were not yet linked. Many samizdat documents from that period had only a very limited circulation inside the Soviet Union and were aimed primarily at a Western audience. This development meant the decline of interest in the broader issues, including environmental problems.

There were some exceptions to that trend. One was the book *The Destruction of Nature* by Soviet environmental screenwriter Zeev (Vladimir) Wolfson. The book was written in the USSR in 1977, smuggled to the West and in 1978 published in Frankfurt under a pseudonym Boris

Komarov. This was one of the first examples of the 'ecocide' narratives that revealed the catastrophic scale of pollution, poaching, mining, logging and erosion. In the chapter 'Our country in reserve', Wolfson polemicised with Solzhenitsyn about the environmental reserves of the Russian North-East, claiming that the ongoing rapacious exploitation of its riches could turn that area into a wasteland within several decades (Komarov, 1978). In his book, Wolfson also mentioned the Kyshtym nuclear disaster of 1957 that had been kept secret in the Soviet Union, although some reports about it appeared abroad, including the statement by emigrated Soviet dissident Zhores Medvedev. The book evoked a wide resonance and received acclaim in the West and was quickly translated into German and English. The *Sunday Times* has even compared it to Rachel Carson's *Silent Spring* (OSA 300–85–12–147/10). In 1979–1982 *Radio Liberty* had a series of broadcasts where chapters of *The Destruction of Nature* were read. Yet, although the book was written inside the USSR (Wolfson emigrated to Israel only in 1981), it is unclear whether it had any circulation in the samizdat format. No samizdat copy of it was preserved in the RFE/RL archives, and I could find no discussion or even mention of this text in any other samizdat publications.

Another exception seemed to be the responses to the river diversion project. Despite the official efforts to silence the debate about it, the project remained a major public concern in the Soviet Union and attracted a fierce criticism from different camps. Apart from scientists, 'village writers' were the famous opponents of the project in official publications, but they also used samizdat channels to communicate their views (Parthé, 1992; Weiner, 1999). Thus, in 1981, one of the stars of Soviet 'village prose', Vasily Belov, wrote an article 'Will Lakes Vozhe and Lacha Save the Caspian' (OSA 300–85–44:34/9), which called for a further scientific evaluation of the river diversion and argued against sacrificing the resources of the North for the development of the South. The article was rejected for publication in the USSR and circulated in samizdat. Other samizdat documents on the topic echoed the same rhetoric, emphasised the value of the northern environment for Russian culture and identity and linked environmental concerns to the protection of cultural heritage (OSA 300–85–9:118/18; 300–85–44:34/9). Yet, as in the case of Wolfson's book, these texts were not produced by the regular samizdat contributors, and the issues they brought up received little attention from other dissidents.

Overall, the large samizdat collections of the RFE/RL Research Institute do not allow us to speak of any sustained interest to environmental problems among Soviet dissidents of the late 1960s–early 1980s. Underground publications touching upon this topic did occasionally appear, and some of them had a strong resonance, especially at the turn of the 1970s, when nonfiction samizdat had wider circulation and broader thematic coverage and when environmental problems were high on the international agenda. However, this did not produce any long-lasting attention or mobilisation around environmental issues in dissident circles and in most cases did not bring up any topics or concerns that had not been already discussed in the official media.

Concluding remarks

So how much did an average Soviet reader hear about global and local environmental change and in what format? The Soviet press did cover a number of environmental issues from a variety of angles, including both optimistic and very critical perspectives. Reporting on local environmental degradation and the risks posed by industrial and urban development started as early as the late 1950s in the more 'elitist' Soviet newspapers and journals but in the Brezhnev era it was taken up by the mass print media. The turn of the 1970s witnessed the rise of interest in global ecological problems and, specifically, climate change.

The publications about local environmental concerns were usually alarmist and normative in tone, calling for immediate action to protect endangered ecosystems. Such texts spoke directly to their readers trying to mobilise their support and often identified specific authorities, organisations or individuals responsible for the situation. The discussion of global issues, including climate change, on the contrary, happened solely through the language of science. Publications on this topic, prepared with the participation of established and highly competent scholars, accurately reported major scientific findings and controversies in the field, presented multiple hypotheses, used a moderate tone and emphasised the need for international scientific cooperation. They did not speak about the differentiated risks and responsibilities of various communities; nor did they try to mobilise the support of their readers or persuade them to change their behaviour, nor generally did they show any interest in individual action. The message that a Soviet reader would get from those publications was that the environmental problems of our planet can only be solved by scientists and governments working together on the global scale and that there is hardly anything an individual could do.

Importantly, most of discussions about the environment happened in the official press. The interest in environmental problems, global or local, among Soviet dissidents of the Brezhnev era was rather weak. Serguei Oushakine once spoke about the 'terrifying mimicry of samizdat', arguing that in the Soviet Union of the 1960s and 1970s, the oppositional discourse echoed the rhetoric of the dominant discourse rather than positioning itself outside of it (Oushakine, 2001: 192). In line with this argument, samizdat texts about the environment do show many parallels with the narratives of the official press, even though in samizdat, free from the tenets of censorship, some topics were amplified and received a sharper articulation. Conservative samizdat generally seems to have had the potential to produce environmental narratives different from those that appeared in the official media but there were no long-lasting debates on environmental topics in dissident circles.

Although the role of censorship in hindering the discussion of many environmental topics is undeniable, large spheres of environmental debate remained 'a little corner of freedom' even in the press. It is plausible that this 'freedom' was exactly the reason why the environment received so little attention in the dissident circles of the 1970s: there was little need to write about it in samizdat because the official press provided sufficient space, and the effect and readership of official publications were certainly much bigger. At the same time, the absence of any significant alternative discourse on global and local environmental change in samizdat might suggest that what limited the (already considerable) range of environmental problems discussed in the official press was not only censorship but also the lack of public concern for the questions beyond it.

Acknowledgements: I would like to thank Marsha Siefert and Istvan Rev for their recommendations and comments on this text.

Note

1 All newspaper and samizdat sources for this chapter were accessed as archival materials and referenced accordingly.

References

Primary sources

Blinken Open Society Archives (OSA).

270–0–1 Radio Liberty (Radio Svoboda) Russian Broadcast Recordings.

300–80–1 Records of RFE/RL Research Institute, Soviet Red Archives, Old Code Subject Files.

300–85–9 Records of RFE/RL Research Institute, Samizdat Archives, Published Samizdat.

300–85–12 Records of RFE/RL Research Institute, Samizdat Archives, Biographical Files.

300–85–12 Records of RFE/RL Research Institute, Samizdat Archives, Unpublished Samizdat: Subject Files.

Secondary sources

Bergman, J. (2009) *Meeting the Demands of Reason: The Life and Thought of Andrei Sakharov*. Cornell University Press, Ithaca.

Breyfogle, N. (2015) 'At the watershed: 1958 and the beginnings of Lake Baikal environmentalism', *The Slavonic and East European Review*, vol 93, no 1, pp. 147–180.

Dowson, J. (1996) *Eco-Nationalism: Anti-Nuclear Activism and National Identity in Russia, Lithuania, and Ukraine*. Duke University Press, Durham.

French, M. (2014) 'Reporting socialism: Soviet journalism and the journalists' union, 1955–1966', PhD thesis, University of Pennsylvania, Philadelphia, US.

Huxtable, S. (2018) 'Making news Soviet: rethinking journalistic professionalism after Stalin, 1953–1970', *Contemporary European History*, vol 27, no 1, pp. 59–84.

Josephson, P. et al. (2013) *An Environmental History of Russia*. Cambridge University Press, Cambridge.

Kelley, D. (1975) *Environmental Problems in the USSR as a Political Issue: Party Adaptation to a New Issue and the Environmental Lobby*, accessed as an archival file in OSA 300–80–1:720/1.

Komarov, B. and Wolfson, Z. (1978) *Unichtozhenie prirody: obostrenie ekologicheskogo krizisa v SSSR*. Posev, Frankfurt/Main.

Oushakine, S. (2001) 'The terrifying mimicry of samizdat', *Public Culture*, vol 13, no 2, pp. 191–214.

Parthé, K. (1992) *Russian Village Prose: The Radiant Past*. Princeton University Press, Princeton.

Show, D. and Oldfield, J. (2016) *The Development of Russian Environmental Thought: Scientific and Geographical Perspectives on the Natural Environment*. Routledge, London.

Weiner, D. (1999) *A Little Corner of Freedom: Russian Nature Protection from Stalin to Gorbachev*. University of California Press, Berkeley.

Wolfe, T. (2005) *Governing Soviet Journalism: The Press and the Socialist Person after Stalin*. Indiana University Press, Bloomington.

4 The environment of an energy giant

Climate discourse framed by 'hydrocarbon culture'

Veli-Pekka Tynkkynen

Introduction

Putin's Russia is highly dependent on fossil fuels and other non-renewable natural resources. This dependence has been discussed previously in the realm of politics and the economy (e.g. Sutela, 2012), but the main argument in this chapter is that this dependence is of a more profound nature, encompassing the spheres of culture and identity, as well. The research question I pose here is: how does the conservative turn experienced in Russian politics and society, and the underlying economic realities of the Putin regime, affect the way the nature and environmental issues are framed in the Russian domestic debate? By debate, I mean the discourses that are constructed, maintained and renewed in Russia's domestic context, thus concerning and targeting the Russian populace. This is an important delimitation, as in the international arena, Russia's communication, for example, concerning climate change, the Arctic and energy trade and cooperation, is very different from the domestic one (e.g. Gritsenko and Tynkkynen, 2018; Tynkkynen and Tynkkynen, 2018). This chapter is based on four articles published previously on climate policies and discourse[1] and on the energy–power nexus[2] in Russia.

The conservative turn under Putin

Vladimir Putin was re-elected to the position of president of the Russian Federation in 2012. Many recent studies support the observation that his re-election marked a further expansion of autocratic elements in Russia's political system (e.g. Gel'man, 2015; Ross, 2015). President Putin's more authoritarian stance since his re-election in 2012, assured already in autumn 2011, is visible both in domestic and foreign policy issues alike. Limiting freedom of expression, limiting gay rights, forcing foreign-funded institutions to register as foreign agents, acting as a decisive military force in the

Syrian civil war, starting the war in Ukraine and subsequently disagreeing with the European Union, and arresting Greenpeace activists in the Arctic all indicate the emphasis on sovereignty rather than international cooperation (cf. Palosaari and Tynkkynen, 2015).

Despite the seemingly drastic changes in Russia's domestic and foreign policy brought forth by the Putin's third term, I argue that there are continuities in the Russian political culture that frame major societal challenges facing Russian regimes. As Kivinen (2002) notes, political decision-making regarding the modernisation agenda of basically all Soviet as well as Russian leaders has allegedly been based on the 'sacred' objectives of science, promoting progress and modernisation and producing economic growth and well-being via expanding industrial production. This consecration has unintended results that are turned into the 'negative sacred' that cannot be addressed in the political and public arenas (Kivinen, 2002: 215–222). The 'negative sacred', especially three such taboos – the demonisation of reality, chaos and consumption – seem pivotal also when trying to understand Russia's stance in global climate politics. The strengthened authoritarian stance presumably indicates that the 'negative sacred' has also strengthened in recent years – there is less effort to justify political decisions for the wider audience, both at home and abroad (e.g. Pomerantsev, 2014; Gel'man, 2016).

Accordingly, Putin's return has not as such contested the policy objectives of modernisation and efficiency that were set during Medvedev's presidency. However, the justifications for these measures have changed and weakened. During Medvedev, energy efficiency and modernisation were justified not only by economic but also by foreign policy and image gains (Henry and MacIntosh Sundstrom, 2012; Korppoo et al., 2015). Since Putin's re-election, the rhetorical need to please foreign audiences has decreased significantly, and the motivation behind modernisation has now a more economic bias, in addition to harsh geopolitical considerations (see Gel'man and Appel, 2015). These developments, according to our analysis, explain also the changed tone on climate change and strengthening link between fossil energy and Russian identity.

Studies such as that of Gustafson (2012) hint that Putin's agenda rests not on diversification of the Russian economy, but on granting the hydrocarbon sector an even greater role in paving the way for Russia's future success. Russian economy and the whole society are firmly tied and dependent on extraction, transport, refining, consumption and export of fossil energy. Fossil energy is central to Russia's economy as more than half of Russia's budget revenue and 70% (in 2014, 54% in 2000) of export is covered by oil, gas and coal; the oil and gas industry alone accounts for a fifth of national GDP.[3] Moreover, the interests behind Russia's national gas programme, run by the parastatal gas giant Gazprom, are at odds with regional interests aiming at energy self-sufficiency via regional renewable sources of energy. In short, Putin's changes in political emphasis have given impetus to the strengthening of Russia's 'hydrocarbon superpower' status (Bouzarovski and Bassin, 2011). An energy superpower is country that is able to influence political choices of other countries via energy exports, by producing dependencies through energy infrastructures (coercive) and economic benefits produced by the energy trade (alluring). Discussion on whether Russia is an energy superpower culminates in the question of how Russia has used energy as a foreign policy tool vis-à-vis its neighbours and the EU, the main customer of Russian energy. Thus, energy wealth and power has been 'productised' into an identity-construction tool, and President Putin can be seen as *the* person responsible for bringing energy assets back to the state and the people from the hands of the oligarchs (e.g. Grib, 2009). Yet, recent studies (e.g. Rutland, 2015; see also Levada Center, 2014) indicate that elites and the public are not particularly attracted, or, more precisely, have an inconsistent and even contradictory attitude, to the idea that hydrocarbons form the fundamental basis of Russia's superpower status or national identity. The elite is aware of the economic problems related to hydrocarbon dependence, and the narrow base of Russia's economy. The people, again, see that due to this dependence – exporting raw materials, importing goods – Russia is easily seen as a developing nation, which does not suit well to the great-power frame in the heart of Russian (national) identity. However, at the same time, the majority of Russians, according to Rutland (2015), see the country as an energy superpower – thus, the weakness of a one-sided economy is turned into a strength. Therefore, in case Putin's entourage wants to strengthen Russia's superpower status based on hydrocarbons, the above-mentioned identity-construction tool leaning on energy and power needs to be used even more aggressively.

At the same time, global hydrocarbon markets have changed significantly during the last couple of years. This change is clear in the gas market, as the 'shale gas revolution' that started initially in the US is reformulating the global gas trade. Production of shale oil is also growing. This has had a negative impact on traded volumes of Russian hydrocarbons, and also on future export prospects (Sharples, 2013). The new energy market situation has been internalised, yet slowly, by Russian leaders and major energy companies during 2011–2012. Dwindling energy export prospects in Europe, coupled with anti-monopoly measures by the EU Commission and price cuts demanded for Russian pipeline gas (Riley, 2012), have possibly motivated the Russian political elite to look for greater export prospects elsewhere, especially in North and Southeast Asia (Bradshaw, 2013), instead of relying on European energy partners, which are institutionally incompatible and demand ethical standards from energy producers. Though, these ethical standards, for example concerning environmental effects of energy production, have not been high on the agenda in the EU's energy policy vis-à-vis Russia. In 2000–2004, the EU-Russia Energy Dialogue had an explicit environmental component to curtail pollution related to oil and gas extraction and transport, but the ecological aims were pushed aside and an economy-driven agenda prevailed from 2004 onwards (European Commission, 2011: 16–19), at the same time as the price of oil and gas increased, and Russia's economy boomed. Thus, I argue that during Putin's third term, the need to pay court to international environmental objectives, due to domestic and international factors, has diminished and Russia's image as a responsible energy producer is of less concern than before. This would leave room for the temptation of downplaying the climate policy objectives and to promote identities leaning on hydrocarbons and fossil energy. More, these two approaches – climate denial and 'hydrocarbon culture' – are but two sides of the same coin: in a nation that sees itself intertwined with the semiotics, materialities and wealth creation of fossil energy (e.g. Kalinin, 2014; Tynkkynen, 2016a), the impetus to act in the forefront of climate politics is a very unlikely choice. In addition, the heightened confrontation between Russia and the west, taking place especially after the outbreak of the wars in Ukraine and Syria, including economic sanctions that target the energy sector, emphasises the Russian need to distinguish itself in all possible ways from the Western-backed agendas.[4] More, as climate change is elementarily linked to the economic base of contemporary Russia and the political power of the ruling regime, i.e. fossil energy, it is no surprise that there is an attraction in this geopolitical situation to define the issue via sovereignty and national identity.

Climate denial – because it's threatening

How climate denial is constructed in Russia?

The internationally prominent community of Russian climate scientists have adopted the international orthodoxy and dismissed the idea that the Russian context could impact their views on climate science, as showed by Wilson Rowe (2012: 712). However, based on our more contemporary study on Russian media (see Tynkkynen and Tynkkynen, 2018), we unfolded three categories of climate change denial – namely conspiracy theories, cherry-picking arguments that fit the denialist narrative and utilisation of outright logical fallacies – that specifically underline the Russian context in terms of political and economic conditions. Thus, Russia's climate change discourse appears nationally specific, especially with regard to denial. It draws on a Russian self-image shaped by geography and resources, Russia's place in the world, and the prevalence and change of historical cultural categories.

The extreme version of the denial discourse promotes the messianic idea that Russia has a special role to play in the global climate system and world history: according to this discourse of climate denial, indicative both of the cherry-picking and logical fallacies categories, Russia needs to 'save the world' from global cooling by emitting more greenhouse gases into the atmosphere. Here, the cooling hypothesis also links to Soviet climate science, thus to the national context, as well, but also to the widely uttered fear of the 1970's academic discourse that we are heading towards a new 'Ice Age'. The milder version underlines that Russia is actually behaving responsibly when it opposes the Western-led green industry conspiracy and declines to compromise global economic growth and, in particular, developing nations' right to modernisation, in the name of climate policy.

A juxtaposition of Russian and international interests regarding climate change is constantly present in our research material on Russian media. International climate policy seems increasingly to be seen as a Western-led hegemonic project aiming to bypass or overrule the sovereignty of Russia. This juxtaposition is supported also by conspiracy-inclined arguments. As our analysis indicates, the denial discourse saws distrust in international climate science and emphasises the positionality and contextual nature of scientific knowledge, in particular by claiming that the West is trying to monopolise climate science and that global climate governance is a western strategy to weaken Russia economically and politically. Parallel arguments, yet with slightly different content, were uttered in the Russian media already in the early 2000s (e.g. Korppoo et al., 2015: 28–29), but it seems that changes in the domestic and foreign political context have created ever more room for them (cf. Laruelle, 2014: 84).

When compared to the Russian climate change discourse during the 2000s (see Wilson Rowe, 2009, 2012; Tynkkynen, 2010), a change can be identified: pessimistic accounts of climate change have gained dominance over the arguments supporting mainstream climate science. Extreme denialists may be no less influential in Russian climate science than before the ratification of the Kyoto Protocol (cf. Laruelle, 2014: 83–84), but as our analysis shows, they seem to have greater opportunities of reaching the public via the media compared to scientists and journalists clinging to the 'mainstream' understanding. The changes experienced recently in Russia's position as hydrocarbon producer and exporter and Russia's foreign and domestic political situation do give further motivation for the political leadership not to oppose climate denial voices in society, if not to support such forces openly.

Why climate denial?

The temporal overlap of the change in the tone of the climate change discourse and Putin's return indicates that the new discourse obviously serves the domestic–political needs of the regime. It might be too far-fetched to argue that the impetus for this qualitative change came after the 2010 drought and fires that devastated Russia and turned many Russian's heads concerning climate change, that is, the need to reduce the threat posed by protestors of the regime, especially as we have not seen much criticism from the public on climate change politics. Yet, the need to downplay criticism toward the regime that has not engaged in climate change mitigation and adaptation is perhaps not fully detached from the fear caused by the protests against Putin's third term in major Russian cities in 2011 and 2012.

On top and as part of the sovereignty argument and direct political interests of Putin's regime, we argue that the material–spatial context of Russia, permeating to the cultural and political spheres, explains at least some of the arguments behind climate denial in Russia. No doubt, one motive for the Russian political and energy actors to oppose mainstream international understanding of climate change, or at least to cast severe doubts on whether climate change is a human-induced process, could draw from both the specific interests of the energy sector in maintaining the status quo in domestic energy policy, as well as the general interests of Putin's regime in reducing the likelihood of criticism by the Russian people toward the hydrocarbon-based political and economic system.

On a theoretical level this denialist strategy leans on a Russian cultural code enabling the use of 'the negative sacral', that is, societal taboos for the benefit of those in power (cf. Kivinen, 2002). In the context of climate denial, three such negative sacreds are of particular interest: demonisation of reality, chaos and consumption. Our analysis indicates that often we see the demonisation of reality constructed especially through the cultivation of conspiracy theories instead of leaning on scientific facts, exemplified in arguments like 'a widely known factor is the interest of (Western) financial giants . . . to engage in trade with greenhouse gas-emissions quotas' (Pavlenko, 2011: 103). Frequently, more exploitation of fossil energy is offered as a cure both for Russia and the developing world, rather than arguing for the decreased extraction of fossil energy (e.g. Channel One, 2009; REN-TV, 2013). This concurs with Jacques (2012) who discusses the general theory of denial: the primary cause for it is because it is serious and threatening for those wishing to maintain power and the accustomed way of life.

Accordingly, the potential and realised chaos caused by climate change is difficult to admit and discuss in the public arena, as one of the *Rossiiskaya Gazeta* (2012) articles claims: 'According to scientists, humans, alas, cannot do anything to avoid such nightmarish forecasts from taking place'. Moreover, the development of production forces, that is, industrial capacity and the following increase in consumption, is viewed as a linear process producing well-being and reducing poverty, as stated in the TV talk show *Gordon Kihot* (Channel One, 2009): ' . . . all possible steps aimed at changing the (global) economy for the benefit of others are taken . . . simultaneously worsening others' possibilities. (This is) based on an academic dispute, nothing more'. This sacral objective is turned into a 'negative sacred', hiding the fact that the extractive nature of the Russian economy ultimately leads to the consumption of the future wealth of the nation through resource depletion and climate change.

Of course, discussions in the state-controlled media do not reflect the attitudes of Russian people, nor do they necessarily predict the moves that Putin's Russia will make in the framework of international climate negotiations (e.g. Korppoo et al., 2015: 44, 47; cf. Smyth and Oates, 2015: 302). Yet, the less the Russian populace is aware of the problems caused by climate change, and especially the less it is alarmed by it, the longer those in power can continue to solidify their power by accumulating wealth through extraction and export of fossil energy and ignore the threats caused by climate change. Promoting ideas of 'undecided climate science', 'non-rational climate agreements' or 'risk-free climate impacts for Russia' fits well together with the interest of the energy industry and Putin's regime to ensure that there is no strong opposition from the grass-root level toward the 'free rider' role of Russia in international climate change mitigation commitments.

For the future of Russia's climate policy, all this comes with major implications. The need for rapid action in the sphere of climate change mitigation may arouse more rejection and denial than agency for change. Because of the 'negative sacred', the potential and realised chaos possibly caused by climate change cannot be discussed. More to the point, as the international climate effort is in Russia often seen as a conspiracy to make profit or limit Russia's sovereignty, the Great Power dimension of national identity makes it difficult to accept the need for forefront climate change mitigation policies and emission cuts. The fossil energy is internalised as a ticket to the class of Great Powers of the world, and it seems that this 'sacred' cannot be questioned any time soon.

Furthermore, referring to the literature on identity construction based on materialities of energy in Russia (e.g. Bassin, 2006; Grib, 2009; Bouzarovski and Bassin, 2011; Rogers, 2012; see also Rutland, 2015), we find that climate denial discourse in Russia could be strategically used to strengthen the national identity constructed on the notion of 'hydrocarbon giant' or 'energy superpower'. International understanding of the problem and especially its internationally agreed solution, including diversification of energy sources away from fossil fuels, are thus pictured in the media material as existential threats to the national identity of Russians. It is this 'hydrocarbon culture' in-the-making we examine next.

'Hydrocarbon culture' – because it's useful

How is 'hydrocarbon culture' constructed in Russia?

Scholars interested in the intertwinement of energy, power and culture in Russia, such as Ilya Kalinin (2014), Douglas Rogers (2012, 2015) and Peter Rutland (2015), have inspired others to engage in similar research with versatile empirical approaches. Our analysis on state-owned gas company Gazprom's advertisements aired on national TV and the web during the early 2010s reveals (see Tynkkynen, 2016a) that the governmental rationality and action, i.e. practiced governmentality by and via Gazprom's Gazifikatsiia Rossii programme derives its power from geographical knowledge and Soviet and post-Soviet imaginaries and from the ability to do 'good' and 'bad'. In this endeavour, the materialities of gas and gas infrastructures are used for both purposes. This produced governmentality is invested with meaning by the existent materialities of hydrocarbons; the pipelines, for example, embody energy security and connectedness to the nation and its resource geography. The physical manifestation of Gazifikatsiia Rossii profoundly affects the construction of the social. Notions of Russia as a territorial superpower and energy superpower are based on the centrality of this materiality. At the same time, this construction lumps together the material specific and nationalistic imagination of energy with universal (neo-liberal) binding goals, such as economic growth and modernisation, and also with particular Russian values, including conservative gender roles. This 'gendered gas' is visible, e.g. in the way women are viewed. They are controlled and fall under the patronage of men, the company and the state, but have some power in their role as healers, consumers and producers of new generations of Russians. These gendered roles date partly from Soviet practices and culture, but they also clearly go hand in hand with the contemporary conservative turn in Russian society and politics. It is no surprise that parastatal Gazprom and the gas industry are viewed as guarantors of this Russian mix of neo-conservative and traditional, patriarchal values. Clearly, gas is a strongly gendered substance and helps to build and maintain a specific form of geographically inspired governmentality.

More, the materialities of gas thus feed into the national identity of Russians as citizens of an energy superpower. This power – projected via international gas pipelines and a military vocabulary – forms the core of the ability to do harm in the domestic arena as well: gas energy, infrastructure and the gas industry are defined and viewed in a manner that underscores the submissive role of individuals and communities. Good Russians 'invite' gas into their community; bad ones leave the community out in the cold by relying on 'non-modern' sources of energy, such as bioenergy.

The production of truths, identity and power in this geography-inspired governmentality includes institutional, physical and administrative mechanisms and knowledge structures. Several discourses, having roots in both the Soviet and post-Soviet nationalistic modernisation ethos, are combined with the spatial and material characteristics of the gas industry to form a compelling narrative where institutional and administrative mechanisms – the gazifikatsiia programme of a parastatal energy company – provide the frame. Furthermore, specific ways of thinking, the understanding of reality, the construction of certain subjectivities and refutation of others, strategic technologies of rule and the values of a specific government are all visible in this endeavour to build a specific 'hydrocarbon culture' (comp. Legg, 2005). Moreover, the rationalities and practices of gazifikatsiia governmentality function in and combine several scales: the subject is tied to territories and the nation through gas, individuals are made responsible for the (bio) security of the population, and even the global is harnessed in legitimising the heavy reliance on gas. Gazprom's Gazifikatsiia Rossii advertisement video shows how the leadership of the company wants gas as a substance and source of energy, the gas industry and the gazifikatsiia programme to be seen by the Russian people.

Why 'hydrocarbon culture'?

The argument presented here is that the way Gazprom links meanings to its tasks as a Russian energy company, and the mentality that can be unfolded by looking the words and deeds of Gazifikasiia Rossii is partly shared by the leadership of the country. Thus, it reflects the efforts of the regime to construct a specific 'hydrocarbon culture'. The overt aim of the video and the governmental thinking behind it is to show how many positive things gas can provide for Russians, but there are subtle hints in the advertisement that gas has the ability to do harm as well. Parastatal Gazprom is there also to control communities and to normalise certain type of identities that reflect the needs of the political elite and those of a nationalistic, aggressive state. As with the climate denial narrative, the construction and maintenance of hydrocarbon culture narrative suits well the needs of the Putin regime, economically and politically highly dependent on oil and gas. The governmental mentality, governmentality visible in Gazprom's advertisements reflects many conservative objectives of the state and the regime, but by far the most important of them is conservative economic policy relying on extraction of natural resources and fossil energy.

Hydrocarbon culture thus can be seen as a tool to prevent popular criticism towards economic policies that resemble those of developing states, and the chosen economic system that relies more and more on the hydrocarbon sector where Russia's role in the global trade is merely a raw material provider, an 'energy-producing appendage' of the West. As Peter Rutland (2015) argues, despite the fact that the majority of Russians consider their country as an energy superpower, most simultaneously oppose the wealth enjoyed by the elite and created by the energy trade, even as many Russians live in factual energy poverty. Therefore, one motivation to produce advertisement videos like this one on Gazifikatsiia Rossii is the need to change this impression and fortify the position of Putin's regime.

This energy culture in-the-making fortifies not only economic and industrial policies, prohibiting them to modernise, but also advocates authoritarian, non-democratic rule and the regime's Great Power ambitions throughout Russia. This is exemplified by the case study we conducted in the Russian Karelia (Tynkkynen, 2016b), where gazifikatsiia was strongly promoted with the help of social infrastructure construction, such as sports halls. Here, discursive (with words) and coercive (with infrastructural objects) governmentality came together in the practices of Gazprom and the Russian state. The amalgamation of energy and sports has enabled the state to practice discursive and coercive power cunningly, as the 'presence' of the state is made concrete through both gas pipelines and visible and spatially extensive sport facilities. Gazprom's all-Russian gas programme and its practices on the local level, as exemplified via the Karelia case study, may be a form of corporate white-washing, but they also advance the Great Power ambitions of Putin's regime in the name of social 'responsibility'. Ultimately, however, Gazprom's sports-oriented social programme aimed to achieve the responsibilisation of individuals – as physically, mentally and *ideologically* fit workers and soldiers – to take care of both the well-being of the self and the nation along with its economy and military might. Its unique form of corporate governmentality can thus be defined as a marriage between the energy superpower ideal and military Great Power identity that are constructed with the help of sports metaphors, value and infrastructures. Thus, sport was utilised to steer energy policies on the local and regional level, as was shown in the Karelian case when the gas programme pushed local bioenergy and energy self-sufficiency goals off the regional agenda. The compelling nationalist narratives manifested in the triangle uniting Russian sports, energy and Great Power status are thus just as important as the mundane energy security objectives, which persuaded Karelian leadership and communities to join Gazifikatsiia Rossii.

Concluding remarks

The Russian hydrocarbon superpower and hydrocarbon culture constructs take a completely different approach to what energy means culturally, socially and economically than we Western observers have become accustomed to. The fact that European consumers have become alienated from fossil energy – how their mundane gas and gasoline are produced, where it comes from, with what social and environmental consequences and how it actually keeps our mobile societies and democracies running – can be seen as a troubling issue. Therefore, the Russian way of constructing an energy culture can also be seen as a more rational way of thinking about society's and the individual's energy dependence than the prevailing Western way. Thus, energy-culture construction efforts such as the Gazifikatsiia Rossii promotional video can serve as a sobering reminder for Western societies of what keeps, in the end, our societies and economies running (comp. Mitchell, 2011). More, contrary to the Western understanding, the Russian people may choose to join the gas infrastructure and voluntarily remain under the patronage of the national monopoly and the federal centre. This positive understanding of patronage certainly has its roots in Soviet history (e.g. Collier, 2011). According to this view, practiced gazifikatsiia governmentality merely mirrors some of the needs of the Russian population.

However, as this 'realist' approach to the issue of how energy and natural resources are intertwined in the social fabric of the society might have its positive sides, the implications of the practices we have observed in contemporary Russia do give room for more worrying thoughts. First, the amalgamation of the needs and rationalities of the fossil energy sector and the domestic and foreign policy interests of the current regime do give grounds to argue that via fossil energy, energy infrastructures and the versatile 'epiphytes' attached to it, such as sport objects, the state has been able to construct and maintain black and white, nationalistic identities. These normalising identities enable the curtailing of possibilities to modernise Russia's economy, suppress political opposition in Russia, and build an illusion that all and everyone is against Russia on the international arena. Second, the emerging energy culture of a fossil giant is trying to monopolise and distort the environmental agenda, transforming it in practice into a social taboo. From regional-level examples, where state energy giants are prohibiting more sustainable energy and environmental policies from developing, to nation-wide propagation of climate denial narrative in the state-controlled media, we see examples of this. Thus, it is unlikely that Russia will take leadership in global climate politics and act in the forefront of efforts to cut emissions. If Russia leads or behaves as a compliant student in global climate policies, it is because of Putin's regime's foreign policy interests, not because there is strong civic opposition among Russians towards Putin's economic, environmental and foreign policies. Therefore, in the near future, another important issue to follow in this field is the way environmental awareness[5] and civic environmental activism is being handled by Putin's regime and its fossil energy entourage.

Notes

1 Tynkkynen, Veli-Pekka and Tynkkynen, Nina (2018). Climate denial revisited: (re)contextualising Russian public discourse on climate change during Putin 2.0', *Europe-Asia Studies*. Online. Skryzhevska, Yelizaveta, Tynkkynen, Veli-Pekka and Leppänen, Simo (2015). Russia's climate policies and local reality, *Polar Geography*, 38 (2), 146–170.

2 Tynkkynen, Veli-Pekka (2016). Energy as power – Gazprom, gas infrastructure, and geo-governmentality in Putin's Russia, *Slavic Review*, 75, 2, 374–395; Tynkkynen, Veli-Pekka (2016). Sports fields and corporate governmentality: Gazprom's all-Russian gas program as energopower, in Koch, N. (ed.), *Critical Geographies of Sport. Space, Power and Sport in Global Perspective*, 75–90. Routledge, Abingdon.

3 See Kurdin, A. (2016). Seminar presentation 'Russian Oil and Gas: Trends and Phenomena to Watch' at a seminar *Russian Oil & Gas: Challenges and Future Developments* organised by the Embassy of Finland in Moscow, October 27, 2016, and Federal State Statistics Service (2015) *Commodity Structure Of Exports Of The Russian Federation*, www.gks.ru/bgd/regl/b15_12/IssWWW.exe/stg/d02/27-08.htm.

4 Naturally, the drastic changes in US climate and environmental policies during the Trump administration blurs the East-West dichotomy on this issue. Trump's US has moved closer to, while China, previously defining itself as a developing nation with limited climate responsibilities, distanced its climate-policy position from, the 'free-rider' Russia. Our research material is, however, collected before 2016.

5 An interesting case to follow is the 'The Year of the Environment 2017' in Russia. Specifically, how state organizations, such as the Russian Geographical Society, in the framework of such projects, try to channel and control civic sentiments and empowerment in the realm of the environment and nature. Judging by the choice of promoted and financed projects under the guise of the Year of the Environment 2017, it looks like there is a very local focus: majority of projects promote household waste and waste water management, as well as aim to curtail industrial pollution. Despite the fact that there is a category of projects called 'The Arctic and Climate', none of the projects addresses climate mitigation per se. This tells that environmental change that is visible for Russians (waste, air pollution) catches also the attention of the regime, but not the global environmental change that will cause much more severe effects for Russians and Russia. This, again, seems to remain in the realm of a taboo for the regime.

References

Bassin, M. (2006) 'Geographies of imperial identity', in D. Lieven (ed.), *The Cambridge History of Russia. Volume II: Imperial Russia, 1689–1917*. Cambridge University Press, New York, pp. 45–64.

Bouzarovski, S. and Bassin, M. (2011) 'Energy and identity: imagining Russia as a hydrocarbon superpower', *Annals of the Association of American Geographers*, vol 101, no 4, pp. 783–794.

Bradshaw, M. (2013) 'The progress and potential of oil and gas exports from Pacific Russia', in S. Oxenstierna and V.-P. Tynkkynen (eds.), *Russian Energy and Security up to 2030*. Routledge, London, pp. 192–212.

Channel One. (2009) 'Gordon Kihot – "Global'noe poteplenie"' [TV programme], 11 December.

Collier, S. (2011) *Post-Soviet Social. Neoliberalism, Social Modernity, Biopolitics.* Princeton University Press, Princeton.

European Commission. (2011) 'Roadmap of the EU-Russia Energy Cooperation until 2050', http://ec.europa.eu/energy/international/russia/doc/20110729_eu_russia_roadmap_2050_report.pdf.

Gel'man, V. (2015) *Authoritarian Russia: Analyzing Post-Soviet Regime Changes.* University of Pittsburgh Press, Pittsburgh, PA.

Gel'man, V. (2016) 'The politics of fear: how Russian rulers counter their rivals', *Russian Politics*, vol 1, no 1, pp. 27–45.

Gel'man, V. and Appel, H. (2015) 'Revisiting Russia's economic model: the shift from development to geopolitics', in *PONARS Policy Memo Series.* George Washington University, Washington.

Grib, N. (2009) *Gazovyi Imperator.* ID 'Kommersant', Moskva.

Gritsenko, D. and Tynkkynen, V.-P. (2018) 'Telling domestic and international policy stories: the case of Russian Arctic policy', in V.-P. Tynkkynen, S. Tabata, D. Gritsenko and M. Goto (eds.), *Russia's Far North: The Contested Energy Frontier.* Routledge, Abingdon, pp. 191–205.

Gustafson, T. (2012) *Wheel of Fortune: The Battle for Oil and Power in Russia.* Harvard University Press, Harvard.

Henry, L. and MacIntosh Sundstrom, L. (2012) 'Russia's climate policy: international bargaining and domestic modernisation', *Europe-Asia Studies* vol 64, no 7, pp. 1297–1322.

Jacques, P. (2012) 'General theory of climate denial', *Global Environmental Politics*, vol 12, no 2, pp. 9–17.

Kalinin, I. (2014) 'Carbon and cultural heritage: the politics of history and the economy of rent', *Baltic Worlds* vol 3, pp. 65–74.

Kivinen, M. (2002) 'Progress and chaos: Russia as a challenge for sociological imagination', in *Kikimora Publications, Series B*: 19. Aleksanteri Institute, Helsinki.

Korppoo, A., Tynkkynen, N. and Hønneland, G. (2015) *Russia and the Politics of International Environmental Regimes. Environmental Encounters or Foreign Policy?* Edward Elgar, Cheltenham.

Laruelle, M. (2014) *Russia's Arctic Strategies and the Future of the Far North.* M.E. Sharpe, Armonk.

Legg, S. (2005) 'Foucault's population geographies: classifications, biopolitics and governmental spaces', *Population, Space and Place* vol 11, no 3, pp. 137–156.

Levada Center. (2014) 'Uchastie Rossii v Bolshoi Vosmerke', www.levada.ru/old/11-04-2014/uchastie-rossii-v-bolshoi-vosmerke.

Mitchell, T. (2011) *Carbon Democracy. Political Power in the Age of Oil.* Verso, London.

Palosaari, T. and Tynkkynen, N. (2015) 'Arctic securitization and climate change', in L.C. Jensen and G. Hønneland (eds.), *Handbook of the Politics of the Arctic*, Edward Elgar, Cheltenham, pp. 165–201.

Pavlenko, V. (2011) *Mify 'Ustoichivogo Razvitiia'. 'Global'noe Poteplenie' ili 'Polzuchii' Global'nyi Perevorot?* OGI, Moskva.

Pomerantsev, P. (2014) *Nothing Is True and Everything Is Possible. The Surreal Heart of the New Russia*. Public Affairs, New York.

REN-TV (2013) 'Territoriia zabluzhdenii s Igorem Prokopenko', No 20. [TV programme], 26 March.

Riley, A. (2012) 'Commission vs. Gazprom: the antitrust clash of the decade?' *CEPS Policy Brief*, 285, 31 October, www.xeps.eu.

Rogers, D. (2012) 'The materiality of the corporation: oil, gas, and corporate social technologies in the remaking of a Russian region', *American Ethnologist* vol 39, no 2, pp. 284–296.

Rogers, D. (2015) *The Depths of Russia: Oil, Power, and Culture after Socialism*. Cornell University Press, Ithaca.

Ross, C. (ed). (2015) Special issue: 'State against civil society: contentious politics and the non-systemic opposition in Russia', *Europe-Asia Studies*, vol 67, no 2.

Rossiiskaya Gazeta (2012) 'Sukhoe budushchee', 5 April, www.rg.ru/2012/04/05/resurs.html

Rutland, P. (2015) 'Petronation? Oil, gas, and national identity in Russia', *Post-Soviet Affairs*, vol 31, no 1, pp. 66–89.

Sharples, J. (2013) 'Russian approaches to energy security and climate change: Russian gas exports to the EU', *Environmental Politics*, vol 22, no 4, pp. 683–700.

Skryzhevska, Ye., Tynkkynen, V.-P. and Leppänen, S. (2015) 'Russia's climate policies and local reality', *Polar Geography*, vol 38, no 2, pp. 146–170.

Smyth, R. and Oates, S. (2015) 'Mind the gaps: media use and mass action in Russia', *Europe-Asia Studies*, vol 67, no 2, pp. 285–305.

Sutela, P. (2012) *The Political Economy of Putin's Russia*. Routledge, Abingdon.

Tynkkynen, N. (2010) 'A great ecological power in global climate policy? Framing climate change as a policy problem in Russian public discussion', *Environmental Politics*, vol 19, no 2, pp. 179–195.

Tynkkynen, V.-P. (2016a) 'Energy as power – Gazprom, gas infrastructure, and geo-governmentality in Putin's Russia', *Slavic Review*, vol 75, no 2, pp. 374–395.

Tynkkynen, V.-P. (2016b) 'Sports fields and corporate governmentality: Gazprom's all-Russian gas program as energopower', in N. Koch (ed.), *Critical Geographies of Sport. Space, Power and Sport in Global Perspective*. Routledge, Abingdon, pp. 75–90.

Tynkkynen, V.-P. and Tynkkynen, N. (2018) 'Climate denial revisited: (re)contextualising Russian public discourse on climate change during Putin 2.0', *Europe-Asia Studies*. Online.

Wilson Rowe, E. (2009) 'Who is to blame? agency, causality, responsibility and the role of experts in Russian framings of global climate change', *Europe-Asia Studies*, vol 61, no 4, pp. 593–619.

Wilson Rowe, E. (2012) 'International science, domestic politics: Russian reception of international climate change assessments', *Environment and Planning D: Society and Space*, vol 30, pp. 711–726.

5 Traditional media and climate change in Russia

A case study of *Izvestiia*

Marianna Poberezhskaya

Introduction

Back in 2004, the attention of the global community was drawn to Russia's climate policy as never before, when Russia's delayed ratification of the Kyoto Protocol finally made it possible to bring the document into force. Russia's prominent position in the global climate regime was partly due to its sizable GHG emissions (fifth largest in the world) and by the withdrawal of the United States from the negotiations (Poberezhskaya, 2016). Whilst in the latest rounds of negotiations attention has moved away from Russia and towards more polluting actors such as China, the USA and India (Korppoo, 2016), it can be argued that a viable climate regime cannot be achieved without Russia's involvement and commitment (Gladun and Ahsan, 2016). Thus, it is important to look at various factors that contribute to and shape Russia's official climate change discourse.

This chapter, in its aim to explore the evolution of the discourse in Russia's traditional media, analyses 668 articles published in Russia's leading newspaper, *Izvestiia*, from 1992 until 2012. It is argued that as during this 20-year period Russia went through major political, economic and social modifications, the media served as a litmus test for understanding how these changes impacted climate change narratives in the country. Even though the number of articles covering climate change has increased over the years, the content of publications has become more polarised with scepticism and a lack of state criticism becoming more apparent. Furthermore, the analysis is not only important for understanding the perception of climate change but also the development of the media's role in modern Russian society.

Russia's climate policy

Russia's geographical characteristics pose a significant dilemma. On the one side, Russia's current political and economic regimes thrive on the abundance of natural resources and in particular from the extensive extraction of fossil fuels. On the other hand, Russia's northern location and territorial spread makes it particularly vulnerable to climate change. According to a Roshydromet report (2017), in 2016 alone, there were 988 cases of extreme weather events in Russia with 380 causing significant economic losses. The report indicates that the increased frequency of these events could be attributed to climate change. Furthermore, it stresses that temperature increases in Russia happen 2.5 times faster than the global rise (temperatures have increased by 0.45°C in the last ten years, whilst in the polar part of the country, it is even worse, with 0.8°C over the same period) (see also Mokhov and Semenov, 2016). Apart from the surge in the frequency of extreme weather events, the negative effects of climate change have already been experienced or will be experienced in the near future throughout the country (Sharmina and Jones, 2015).

Despite its geographical vulnerability, over the years, Russian climate change policy has been conditioned by a range of political and economic factors. For example, economic development has been prioritised (Henry, 2010), with the state's support persistently being allocated to the polluting fossil fuel industry. Even Russia's infamous ratification of the Kyoto Protocol took place in the context of political bargaining and international negotiations (e.g. EU support of Russia's WTO membership). It should also be noted that agreement did not require Russia to make any economic sacrifices due to the industrial collapse experienced after the USSR's dissolution. This had resulted in Russia's GHG emissions drastically dropping by the late 1990s compared to the 1990 baseline year, meaning Russia would meet its Kyoto targets without any additional effort. However, this provided an opportunity for domestic political actors to present the country as an 'environmental hero', leading the way in the global fight against climate change (Poberezhskaya, 2016).

In fact, this narrative of Russia's great climate mitigation contribution has been persistently underlined by both Dmitry Medvedev and Vladimir Putin (despite the fact that the latter has, on a number of occasions, expressed his scepticism of the anthropogenic character of climate change or its negative effect on Russia) (ibid, see also Chapter 4). The political climate change narrative experienced a pronounced change in 2009, when Dmitry Medvedev attended the Copenhagen Conference, and a number of national documents were accepted (e.g. the Climate Doctrine (2009) and Climate Doctrine implementation plan (2011)). In 2013, Russia's national climate policy was strengthened by the Presidential Decree 'On the reduction of GHG emissions' (2013) and Putin's announcement at the COP-21 of Russia's commitments to cutting carbon emissions by 25 percent by 2030 (of the baseline year 1990). The latest statement made by the Special Envoy of the President of Russia for Climate Affairs, Alexander Bedritsky reinforced these obligations, whilst the importance of counting the influence of Russia's boreal forest has been emphasised once again (Bedritsky, 2017). There also has been a tendency to emphasise the positive achievements in Russia's domestic climate policy, for example, 2017's Russian Climate Week, in which a number of Russian businesses took part, or the creation of the Climate Partnership of Russia, which in its own words, 'consolidates the efforts of Russian business to mitigate environmental impacts and help prevent climate change' (climatepartners.ru, 2017). The latter includes the world's second-largest aluminium company, Rusal.

More recently, as the conflict in the Eastern Ukraine unfolded, some positive advances in Russia's climate policy have been reversed, as budgets have had to be adjusted due to the economic crises that have followed (Davydova, 2015). Interestingly, the international tensions caused by the prolonged conflict in Ukraine have penetrated Russia's official discourse on climate change:

> At the same time, we believe that the politicization of socio-economic cooperation, including its climatic aspects, the imposition of economic sanctions on a number of countries that are Parties to the UNFCCC, hinders the successful implementation of measures to reduce greenhouse gas emissions in these countries and their climate-resilient development. Further implementation of the sanctions policy with regard to a number of countries will call into question the joint achievement by the countries of the objectives of the Paris Agreement.
>
> (Bedritsky, 2017)

Russia's positive advances in domestic climate policy have been met with a bit of scepticism. Both scholars and activists point out that such national documents as the Climate Doctrine or its Implementation Plan 'do not contain effective tools to reduce GHG emissions' (Gladun and Ahsan, 2016: 27), and Russia's 'ambitious' commitments to cut emissions by 25% to 1990 levels can result in only an 11% decrease if forestry is considered (climateactiontracker.org, 2017). Korppoo and Kokorin (2017) argue that Russia's domestic mitigation policies remain quite weak, handicapped by loopholes in the legislative documentation as well as numerous bureaucratic barriers. Furthermore, Russia's other economic and political strategies openly contradict its climate change-related policies (e.g. state support of the fossil fuel industry) (Sharmina and Jones, 2015). However, Russia's involvement in the Ukrainian conflict and the subsequent economic decline, mean that once again Russia's climate policy has unexpectedly benefitted as a result of the reduced rate of growth in its economy – which has led to a slowing down of GHG emissions increase (ibid).

A number of research studies have demonstrated that whilst Russia's climate policy has been influenced by a range of factors, the state's leaders (in particular Vladimir Putin) have been 'the most influential actor[s] in the decision-making process' (Korppoo, 2016: 644). Despite all the obvious downsides of this predicament, in theory, it has one potentially positive consequence which was pointed out by Sharmina et al. (2013: 389), who argue that 'the semi-authoritarian policy regime in Russia can more readily impose climate-related policies on the population, with a historically passive civil society further facilitating the top-down approach'. Finally, an overview of Russia's official climate discourse would be incomplete without mentioning its relatively low level of concern and awareness of climate change amongst the public (Sharmina et al., 2013). Even the population of the highly vulnerable geographical locations in the north of the country, who are able to observe and experience the adverse consequences of climate change directly, remain mostly sceptical of the anthropogenic nature of climate change (Graybill, 2013). Arguably, one of the reasons behind this is the restricted media coverage and close connections between media and political discourses, which will be explored further in this chapter.

Climate discourse and Russian media

Over the last couple of decades, Russian mass media went through substantial transformations. As Erzikova and Lowrey (2014: 36) eloquently summarise, 'a number of socio-political developments have led Russian mass media from being over-politicised in the 1990s to becoming politically apathetic in the 2000s'. Indeed, towards the end of the USSR's existence and during the emergence of the new Russian Federation, mass media came closest to becoming a 'fourth estate' and playing an important role in political processes (Grabel'nikov, 2001). However, in the following years some media outlets changed ownership meaning that instead of exercising power, they became a powerful tool in someone else's hands (Azhgikhina, 2007). In the early 2000s, Russia's political leadership changed and was accompanied by the centralisation of major media outlets, which coincided with Russian media being classed as 'unfree' by the Freedom House after enjoying years of 'partial freedom' (Toepfl, 2013). Whilst modern Russian journalism is facing a number of challenging barriers with certain topics being of a particularly sensitive nature and, therefore, they are either avoided by the media as a whole or cause various problems for specific media outlets (the degree of damage depends on the sensitivity of the topic), climate change (despite its controversy) has not caused extreme confrontation and censorship by the state (Poberezhskaya, 2016).

In fact, Russian media have followed some global patterns in climate communication where journalists are seen as a key channel in making complex issues of climate change visible to the wider public and intentionally or unintentionally playing an important role in forming people's opinions about climate polices and affecting their behaviour by bridging scientific, political and environmentalist communities (e.g. Nelkin, 1987; Bell, 1994; Carvalho 2007; Boyce and Lewis, 2009; Olausson, 2009; Boykoff, 2012). There are also, of course, some country-specific particularities regarding climate change communication. For instance, scholars agree that in the Russian case, the state's influence over the media has also penetrated the climate change discourse. Whilst there are no documented cases of censoring the climate change topic, various media analyses have shown a correlation between changes in the state policy on climate change and the quality and quantity of media coverage on the topic (Wilson Rowe, 2009; Tynkkynen, 2010; Yagodin, 2010; Poberezhskaya, 2014).

One of the distinct features of climate change coverage in Russia is the relative absence and unpopularity of the topic, which can be referred to as 'climate science' (Poberezhskaya, 2016). Indeed, Russia falls far behind in its coverage of not only the major Western powers but also countries with similar or lower levels of economic and political development. Even social media, which could be seen as an alternative platform for advancing media discussion, have been mostly overlooked in Russia or often serve as a channel for re-enforcing conspiratorial and sceptical climate change-related discussion (Poberezhskaya, 2017). A more detailed and systematic analysis of media coverage conducted by Boussalis et al. (2016) also shows when climate change is more likely to appear on the pages of Russian newspapers and within what context. For instance, they determined the importance of economic conditions (which was measured by Russian inflation levels). When the economic situation worsens, Russian media have mostly ignored climate change or discussed it more within the context of geopolitical problems (e.g. what Russia needs to do not to lose a competitive advantage) and less in the context of science and international commitments. Building up on the existing body of literature, this chapter, by using the example of one leading newspaper, demonstrates what exactly is happening with climate coverage during its peaks and troughs.

Methodological considerations

This chapter concentrates on the coverage of climate change in one of the most popular and authoritative Russian newspapers *Izvestiia*. Whilst *Izvestiia* does not have the largest circulation in the country (Boussalis et al., 2016), it can be looked at as an interesting and rather typical example of how the role of the social-political newspaper has adapted throughout the decades after the dissolution of the Soviet Union. *Izvestiia* was founded in

1917 and for over seven decades served as one of the main media platforms of the Communist Party. However, it also developed popularity among Soviet intellectuals and academics. When the new state of Russia was created, *Izvestiia* went through a series of modifications by first showing its independence and playing important roles in the state's policy, but towards the end of the 1990s, *Izvestiia* had to sacrifice its freedom and accept financial support from an oil giant LUKoil (Voltmer, 2000). Later, it was owned by Gazprom and then sold to another gas company SOGAZ and 'media baron' Yurii Koval'chuk.

For this analysis, the keyword search (all grammatical variations of climate change, global warming and greenhouse effect) was conducted throughout all publications by *Izvestiia* from 1992 until 2012. After the manual examination of the collected articles (making sure that articles with irrelevant use of climate were eliminated), 668 publications were identified. The goal of this research study is to explore the themes in the coverage of climate change in the traditional mass media in Russia. The methodology was inspired by the discourse analysis suggested by Maarten Hajer (1995), who looks at how various perceptions of environmental problems evolved over time and influenced environmental politics. Collecting articles over a 20-year period allowed the author to conduct a 'historical-diachronic' analysis of climate change discourse (Carvalho, 2008). Therefore, we can see how certain themes evolved and developed as Russia was going through changes in its political regime (the timeframe covers two of Yeltsin's presidential

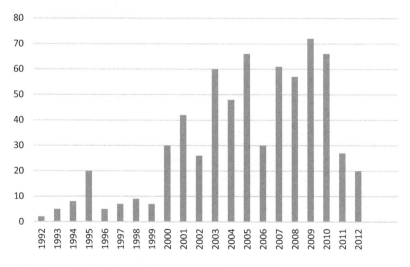

Figure 5.1 Izvestiia climate change coverage, 1992–2012

terms, two of Putin's and one of Medvedev's presidential terms) and economic situation (starting with the economic collapse of the early 1990s and then moving towards the more stable situation with the country's growing reliance on fossil fuel exports).

As can be seen from Figure 5.1, *Izvestiia* coverage followed similar patterns to other Russian media (see Boussalis et al., 2016). Whilst, relatively speaking, it has been quantitatively behind Western media, the peaks and downfalls in coverage seem to be impacted by external events (e.g. political and economic problems in the 1990s or various key climate change political events such as the Copenhagen Conference in 2009). However, a more detailed analysis is required in order to establish the evolution of certain topics and qualitative change in publications.

Exploring themes in *Izvestiia's* coverage of climate change

The following analysis has been divided around four time periods (1992–1999; 2000–2004; 2005–2008 and 2009–2012). This delimitation has been motivated by the major changes in Russia's politics or economics (such as changes at the executive level). However, it should be admitted that there was some degree of arbitrary judgement. For instance, the first period is substantially longer and covers eight years due to a very low number of articles published during this time and a lack of substantial changes in the governance of the country. The next two periods reflect two terms of president Putin in office, whilst the last one covers Medvedev's reign.

1992–1999

In the eight years after Russian independence, *Izvestiia* published 63 articles with some discussion of climate change. The limited coverage is understandable as the country was in complete economic and political disarray with major official institutions (including the ones in charge of the environment) experiencing rapid and sometimes numerous changes of function and structure. As was mentioned above, media has been actively involved in the political process at the start of the 1990s and towards the end of that decade, many media found themselves in the hands of oligarchs (often with strong connections to the energy industry). *Izvestiia* was no exception to this dynamic (Voltmer, 2000).

Despite the very limited coverage, the quality of this now historic discussion is interesting as almost half of all publications (n=29) explicitly discuss the negative consequences of climate change. And the majority of publications (n=39) refer to the anthropogenic character of the observed climatic changes (whilst only five articles refer to natural climate change and others do not specify the reason for the observed climatic changes). For example, the very first article describes the health threatening air pollution in Mexico caused by industrial GHG emissions, which are 'a main culprit of global warming' (Kovalev, 1992). The range of negative consequences of climate change includes threats to human health, a reduction in biodiversity and the undermining of national and economic security (the latter could be further impacted by increased numbers of environmental migrants or restrictions on the energy sector due to global mitigation policy) (for example, see Kovalev, 1993; Bovkun, 1995; Platkovskii, 1996).

The Kyoto Conference in 1997 provoked a new theme in climate change-related publications – discussion of international negotiations and how the global climate regime might affect Russia. Overall, these articles talk about Russia's carbon advantage (its rapid drop in GHG emissions), which has the potential to translate into economic benefits. However, the articles do not blindly praise Russia's position in the global climate regime but sometimes offer an acute critique of the politicians and bureaucrats involved. As an example, the article titled 'Russia will become the largest air trader' (Zhuravlev and Leskov, 1998) begins with a sensationalist statement about how Russia might earn $3 billion from carbon trade. Further on the discussion turns quite critical of the global climate regime which, instead of actively reducing the emissions of developed countries, allows them to buy quotas and continue to pollute, whilst people in charge of the carbon trade in Russia do not care about the global rise in emissions stipulated by these transactions as for them 'money has no smell'[1]. Over this period, only two articles clearly express scepticism towards the nature and existence of climate change (e.g. Bateneva, 1997).

2000–2004

Over the next five years, a gradual increase in the newspaper's climate change coverage can be observed. Trends which have been detected in the first period continued. Climate is more often discussed within the context of its negative influence on nature, human health or the national economy. Its anthropogenic nature is mentioned in half of the articles (n=97) and with only a small proportion of the articles discussing other causes of global warming (and the rest not referring to the causes).

This period is characterised by journalists' attention to the intricacies of the Kyoto ratification process. The whole process was a perfect combination of diplomatic drama, economic controversy and an opportunity to critique Russia's historical opponent – the US. A good example would be an article with a telling title 'America does not care about the planet' (Pimenov, 2001). After explaining how the US is stopping the Kyoto Protocol from coming into force, the author lists the opinions of politicians and scientists from around the world with strong accusations directed towards the US: 'Bush's policy is called irresponsible and scandalous', 'with the US the Kyoto Protocol turns into a dead paper', 'these actions will lead to climate catastrophe' (ibid). The author of the article shows his agreement with all these attributes and notices that this 'unhappy story' demonstrates the truthfulness of the proverb that 'America is above everyone'.[2] It should be noted that the Kyoto discussion has also allowed journalists to mention the disagreements within the Russian scientific and political community. For example, a number of articles are dedicated to disputes involving Russia's infamous antagonist of the Kyoto ratification, Andrey Illarionov, who was economic policy advisor to the president at that time, and who likened the document to a 'plague' (Leskov, 2004), meaning that it will be as destructive. However, what is interesting despite the high status of the Protocol's antagonist is that the journalists did not shy away from questioning his credibility: 'can one learn the whole [climate] science in six months?' (Leskov, 2004). Furthermore, the journalists' criticism did not spare the leader of the country [V. Putin], whose statement during the Moscow conference on climate change has surprised many with its 'careless attitude to the problem of climate change' (*Izvestiia*, 2003). Another interesting characteristic of the Kyoto coverage of that time concerns the birth or a re-establishment of Russia's environmental leadership status. Perhaps the only time in the history of climate change global negotiations that Russia had a tangible justification for its importance in climate governance (e.g. 'They cannot wait for us to join. The Kyoto Protocol's fate depends on Russia's decision' (Valstrem, 2003)).

This period is also interesting because of a relative spike in coverage in 2003, which to some extent can be explained by an unprecedented event – the World Climate Change Conference in Moscow.[3] The conference was initiated by Vladimir Putin, who presented it as an opportunity to gather information on the complicated subject in order to form a decision on the ratification of the Kyoto Protocol. The conference received mixed reviews from the international scientific community, and as mentioned above, Putin's controversial opening remarks were criticised by both foreign and national media (e.g. *Izvestiia*, 2003). Nevertheless, from the environmental communication point of view, the event has created a good opportunity for initiating or stimulating public discussion.

2005–2008

The following timeframe saw a similar quantity of articles dedicated to the topic of climate change (n=214). Almost half of the publications had some references to or indications of the anthropogenic character of climate change (n=90), whilst the other half (n=114) did not make the nature of climate change clear (with a small number of articles referring to the natural processes of climatic modifications). The topics of the publications remain quite diverse throughout the studied period. However, there are certain trends in the quality of the articles. For instance, as climate change becomes a regular item of international political summits, almost a quarter of all texts (n=43) mention it in passing as one of the items of negotiations or discussions. Whilst this does not provide an elaborate account of the problem, it shows its imminent importance and the level of its potential impact. The most popular theme remains the discussion of the negative consequences of climate change (n=61), which once again outlines a range of impacts, starting with the increased number of extreme weather events and finishing with the rising number of health problems.

What is interesting during this period is that we can see a slightly larger number of articles discussing the positive effect of climate change for Russia and also for the Arctic which is supposed to allow freer access to natural resources. For example, a publication titled 'Great Arctic divide. Global warming opens up unprecedented economic opportunities' (Krauss et al., 2005) lists all benefits which melting ice will bring to countries with access to the Arctic's resources (energy resources, fishing, the Northern Sea Route). This discussion coincides with the renewed interest of the Russian state in Arctic affairs (Khrushcheva and Poberezhskaya, 2016). Similarly, as with the previous timeframe, this period also demonstrates a slow growth of scepticism in the coverage. If between 2000 and 2004, 12% of publications had some sceptical sentiments, in the 2005–2008 period, 14% (n=29) of articles either stated that the observable changes in weather patterns have nothing to do with global warming, or doubted the anthropogenic nature of climate change. For example, in the publication 'There is no change on the atmospheric front', the author, whilst outlining the conventional position of the 'UN experts', also provides an elaborate discussion on the potential use of the 'climate weapon' (Obraztsov, 2007).

2009–2012

In this period, two very significant climate-related events took place. Firstly, the Copenhagen Conference in 2009 witnessed a very clear statement from Russia's head of state that confirmed Russia's desire to participate in the global climate regime (see more in Poberezhskaya, 2016). Secondly, in 2010 Russia fell victim to extreme weather events (a heat wave which provoked violent fires and droughts throughout the country resulting in severe human and economic losses). However, despite the positive changes in Russia's national climate policy and negative changes in its climate, *Izvestiia* started to more often refer to the sceptical frame in their coverage with almost 30% of articles in this period (n=50) doubting the nature of global warming, its negative impact or its existence at all. The change in sceptical discussion was not only manifested through a quantitative increase but also by a move towards a more assertive tone for this type of publication. For instance, in 2009, *Izvestiia* published an article titled 'Crisis has stopped financing of "global warming"' (Obraztsov, 2009) which has taken the discussion to the next level. The publication begins with unapologetic statements that volcanoes and forest fires should be blamed for global warming which has been proved by two 'large scale research projects' with 'sensationalist' results. What is more interesting is that after this publication a number of other articles (on various climate-related topics) have referred to this debunking of the 'climate change myth' in *Izvestiia*. For instance, the publication about a new 'environmental car' project (made out of recycled materials and run on bio-fuel) does not just question how realistic this project is but also highlights that the 'role of humans in global warming turned out to be a myth (which *Izvestiia* published about yesterday)' (Streltsov, 2009). Then we can see the re-appearance of the casual mentioning of the absurd nature of the 'global warming theory'. For example, in the discussion on real and pseudo-scientists, the author praises a late Soviet climatologist who 'was not afraid to make a statement against another international scam named "global warming"' (Melikhov, 2010).

Whilst a substantial number of articles (n=31) still report on the negative impacts of climate change (though it is not always linked with an anthropogenic contribution), we can see the continuation of the theme discussing the positive impact of climate change on Arctic development and how it will bring benefits to Russia's economy. The analysis has also shown that now the climate change topic cements itself within political discourse with one third of all publications either discussing the global climate regime (conference, negotiations and agreements) or mention it among other political items or global threats.

Concluding remarks

In 1992, just a few months after the collapse of the USSR, at a time when the country had very little sense of its economic and political future, *Izvestiia*'s journalists wrote only two articles on climate change. But both publications firmly stated the anthropogenic character of climate change and its devastating impact on humans' lives, where one journalist emotionally writes, 'this way nature demanded: do not rape me' (Kovalev, 1992) and another provides an elaborate explanation of 'warming which makes you shiver' (*Izvestiia*, 1992).

Two decades later, worldwide we are witnessing the advancement of the climate change discussion. The IPCC published its fourth Assessment Report in which it stressed that global warming is 'unequivocal' and that it is 'very likely' that the observed changes are due to human activity (IPCC, 2007). At the same time, Russia, after realising a range of economic benefits from the development of energy efficiency programmes and estimating losses from its climate vulnerability, managed to adopt a number of documents at the national level to mitigate the advancement of climate change. Even though many still doubt the usefulness of these national documents, it was seen as a certain change in official rhetoric. Furthermore, the global community has slowly moved towards a consensus on climate change mitigation and adaptation by hosting the 2012 United Nations Climate Change Conference in the Middle East (Doha, Qatar). *Izvestiia* started to mention and discuss climate change more often in the 2000s (with the number of publications peaking in 2009 – which was in line with the global trend provoked by the highly publicised Copenhagen conference), but in 2012 it published only 20 articles – the lowest number since 2000. Whereas numbers alone are quite interesting for our observations of the development of climate change discourse, a more striking pattern has been observed within the qualitative change.

Although the global scientific community (including their Russian colleagues) have become more confident in climate science, the discussion in *Izvestiia* becomes more sceptical. If in the 1990s and early 2000s, sceptical themes were practically absent, in the last four years of the studied period, almost one-third of all publications refute either the anthropogenic character of climate change or its negative impact. Moreover, if we look at all the published articles over the 20-year period, the sceptical sentiments are still marginal (n=109, 16%), however, the majority of articles do not have any references to the nature of observed climate change (60% of all publications). This means that the author can provide a very accurate account of climate change impacts and raise a number of important problems, but would fail to allocate the blame for it, leaving the reader with no choice but to accept natural abnormalities as something unavoidable. This evolution of scepticism could be seen as puzzling if one expects media discourse to follow scientific discourse, though, if we look at the arguments developed in other chapters of this book, Russia's scepticism is rather predictable due to its political and economic interests. At the same time, Kokorin (2017) in his explanation of the roots of Russia's climate scepticism explores people's suspicion of information coming from scientific or political authorities as the country went through decades of open propaganda and adjustments of facts (when needed). Therefore, people tend to question any information which comes from 'above' or as Kokorin states: 'there is less faith and more scepticism' (2017: 106). Besides, if we isolate the Russian scientific discussion from the global one, we can see that scepticism is still present there (arguably more than in Western countries). In this regard, it is interesting to look at the study conducted by Dronin and Bychkova (2017), who argue that present scepticism in the Russian scientific community is due to the Soviet heritage of a different approach to studying nature.

Of course, one can also look at the changes in the newspaper's coverage, its ownership structure and the evolution or devolution of its political role in Russia. This links with another observed trend, which is that the first part of the analysed timeframe contains at least some criticism of the political regime and its approach to the environment and economy, yet after 2004 we can see only a couple of similar discussions. Instead, the newspaper reiterates the state's expectation of economic benefits from taking part in international negotiations on climate change or utilising the 'warmer' Arctic. This again coincides with more general developments in Russia's media sphere, which after a decade of relative freedom were gradually driven towards giving up their watchdog function (Lipman, 2009).

Climate change is still a relatively low-key problem in Russian official and public discourses, meaning that it does not monopolises the discussions or does not enter the list of politically sensitive topics. However, it seems to have fallen victim of the overall propensity to avoid questioning the state's policy or initiating a discussion of Russia's controversial economic and environmental policies.

Notes

1 This is a common Russian phrase which means regardless where money is coming from, they still have the same value.
2 This proverb refers to the Russia's perception of US exceptionalism that goes hand in hand with its arrogance and little concern for anyone else's well-being.
3 This should not be confused with the World Climate Conferences organised by the World Meteorological Organisation or the United Nations Climate Change Conferences (officially known as UNFCCC Conferences of the Parties).

References

Azhgikhina, N. (2007) 'The struggle for press freedom in Russia: reflections of a Russian journalist', *Europe-Asia Studies*, vol 59, no 8, pp. 1245–1262.
Bateneva, T. (1997) 'Morozy russkim krasavitsam k litsu', *Izvestiia*, 23 October.
Bedritsky, A. (2017) 'Zaiavlenie spetspredstavitelia Prezidenta po voprosam klimata Aleksandra Bedritskogo', www.kremlin.ru/events/administration/56013.
Bell, A. (1994) 'Climate of opinion: public and media discourse on the global environment', *Discourse & Society*, vol 5, no 1, pp. 33–64.

Boussalis, C., Coan, T. and Poberezhskaya, M. (2016) 'Measuring and modeling Russian newspaper coverage of climate change', *Global Environmental Change*, vol 41, pp. 99–110.

Bovkun, E. (1995) 'Pristupy astmy i chesotki ot tsvetochnoi pyli', *Izvestiia*, 29 June.

Boyce, T. and Lewis, J. (eds). (2009) *Climate Change and the Media*. Peter Lang, New York.

Boykoff, J. (2012) 'US media coverage of the Cancun climate change conference', *Political Science & Politics*, vol 45, no 2, pp. 251–258.

Carvalho, A. (2007) 'Ideological cultures and media discourses on scientific knowledge: re-reading news on climate change', *Public Understanding of Science*, vol 16, no 2, pp. 223–243.

Carvalho, A. (2008) 'Media(ted) discourse and society', *Journalism Studies*, vol 9, no 2, pp. 161–177.

Climate Doctrine of the Russian Federation. (2009) http://archive.kremlin.ru/eng/text/docs/2009/12/223509.shtml.

Climateactiontracker.org. (2017) 'Russian federation – climate action tracker', http://climateactiontracker.org/countries/russianfederation.html.

Climatepartners.ru. (2017) 'Climate partnership of Russia', http://climatepartners.ru/en/.

Comprehensive Implementation Plan of the Climate Doctrine of the Russian Federation for the period up to 2020. (25 April 2011) Directive No. 730-p of the Government of the Russian Federation.

Davydova, A. (2015) 'Russia's forest overlooked in climate change fight', *Thomson Reuters Foundation*, 15 January, www.trust.org/item/20150115092042-mtqjn/?source=jtOtherNews1.

Dronin, N. and Bychkova, A. (2017) 'Perceptions of American and Russian environmental scientists of today's key environmental issues: a comparative analysis', *Environment, Development and Sustainability*, https://doi.org/10.1007/s10668-017-9979-8.

Erzikova, E. and Lowrey, W. (2014) 'Preventive journalism as a means of controlling regional media in Russia', *Global Media and Communication*, vol 10, no 1, pp. 35–52.

Gladun, E. and Ahsan, D. (2016) 'BRICS countries' political and legal participation in the global climate change agenda', *BRICS Law Journal*, vol 3, no 3, pp. 8–42.

Grabel'nikov, A. (2001) *Rabota Zhurnalista v Presse*. Rip-holding, Moscow.

Graybill, J. (2013) 'Imagining resilience: situating perceptions and emotions about climate change on Kamchatka, Russia', *GeoJournal*, vol 78, no 5, pp. 817–832.

Hajer, M. (1995) *The Politics of Environmental Discourse: Ecological Modernisation and the Policy Process*. Oxford University Press, Oxford.

Henry, L. (2010) 'Between transnationalism and state power: the development of Russia's post-Soviet environmental movement', *Environmental Politics*, vol 19, pp. 756–781.

IPCC (Intergovernmental Panel on Climate Change). (2007) 'Climate change 2007: synthesis report', www.ipcc.ch/pdf/assessment-report/ar4/syr/ar4_syr.pdf.

Izvestiia (1992) 'Poteplenie, ot kotorogo brosaet v kholod', *Izvestiia*, 22 May.

Izvestiia (2003) 'Kholodnaia voina', *Izvestiia*, 15 October.

Khrushcheva, O. and Poberezhskaya, M. (2016) 'The Arctic in the political discourse of Russian leaders: the national pride and economic ambitions', *East European Politics*, vol 32, no 4, pp. 547–566.

Kokorin, A. (2017) 'Analiz problem skepticheskogo otnosheniia k antropogennym prichinam izmeneniia klimata', *Izpol'zovanie i okhrana prirodnykh resursov v Rossii*, vol 14, pp. 105–109.

Korppoo, A. (2016) 'Who is driving Russian climate policy? applying and adjusting veto players theory to a non-democracy', *International Environmental Agreements*, vol 16, pp. 639–653.

Korppoo, A. and Kokorin, A. (2017) 'Russia's 2020 GHG emissions target: emission trends and implementation', Climate Policy, vol 17, no 2, pp113–130

Kovalev, I. (1992) 'Gorod chut' ne zadokhnulsia', *Izvestiia*, 08 May.

Kovalev, I. (1993) 'Navodnenie v Bandgladeshe – delo privychnoe', *Izvestiia*, 11 June.

Krauss, K., Mayers, S.L., Revkin, A. and Romero, S. (2005) 'Velikiy peredel Arktiki. Global'noe poteplenie otkryvaet nevidannye ekonomicheskie vozmozhnosti', *Izvestiia*, 21 October.

Leskov, S. (2004) 'Politekologiia. Rossiia prisoediniaetsia k Kiotskomu Protokolu', *Izvestiia*, 01 October.

Lipman, M. (2009) 'Media manipulation and political control in Russia', Chatham House Report, www.chathamhouse.org/publications/papers/view/108964.

Melikhov, A. (2010) 'Zasukha v golovakh', *Izvestiia*, 25 August.

Mokhov, I. and Semenov, V. (2016) 'Weather and climate anomalies in Russian regions related to global climate change', *Russian Meteorology and Hydrology*, vol 41, no 2, pp. 84–92.

Nelkin, D. (1987) *Selling Science: How the Press Covers Science and Technology*. Freeman, New York.

Obraztsov, P. (2007) 'Na atmosfernom fronte bez peremen', *Izvestiia*, 22 February.

Obraztsov, P. (2009) 'Krizis pokonchil s finansirovaniem "global'nogo potepleniia"', *Izvestiia*, 27 April.

Olausson, U. (2009) 'Global warming – global responsibility? media frames of collective action and scientific certainty', *Public Understanding of Science*, vol 18, pp. 421–436.

Pimenov, A. (2001) 'Ameriku planeta ne volnuet', *Izvestiia*, 31 March.

Platkovskii, A. (1996) 'Kitai dvizhetsia ne na sever, a na iug', *Izvestiia*, 23 July.

Poberezhskaya, M. (2014) 'Media coverage of climate change in Russia: governmental bias and climate silence', *Public Understanding of Science*, vol 24, no 1, pp. 96–111.

Poberezhskaya, M. (2016) *Communicating Climate Change in Russia: State and Propaganda*. Routledge, Abingdon.

Poberezhskaya, M. (2017) 'Blogging about climate change in Russia: activism, scepticism and conspiracies', *Environmental Communication*, https://doi.org/10.1080/17524032.2017.1308406.

President of the Russian Federation. (2013) Decree no. 752 "On Reduction of Greenhouse Gas Emissions", 30 September.

Roshydromet. (2017) 'A report on climate features on the territory of the Russian federation in 2016', http://cc.voeikovmgo.ru/images/dokumenty/2017/doc2016. pdf.

Sharmina, M., Anderson, K. and Bows-Larkin, A. (2013) 'Climate change regional review: Russia', *Wiley Interdisciplinary Reviews: Climate Change*, vol 4, no 5, pp. 373–396.

Sharmina, M. and Jones, C. (2015) 'Discounting the future of climate change in Russia, open democracy', www.opendemocracy.net/od-russia/maria-sharmina-christopher-jones/ discounting-future-of-climate-change-in-russia.

Streltsov, E. (2009) 'V polnoch' tykva prevrashchaetsia . . . v avtomobil'!', *Izvestiia*, 28 April.

Toepfl, F. (2013) 'Why do pluralistic media systems emerge? comparing media change in the Czech republic and in Russia after the collapse of communism', *Global Media and Communication*, vol 9, no 3, pp. 239–256.

Tynkkynen, N. (2010) 'A great ecological power in global climate policy? framing climate change as a policy problem in Russian public discussion', *Environmental Politics*, vol 19, pp. 179–195.

Valstrem, M. (2003) 'Nas zhdut s neterpeniem. Ot resheniia Rossii zavisit sud'ba Kiotskogo protocola', *Izvestiia*, 28 June.

Voltmer, K. (2000) 'Constructing political reality in Russia', *European Journal of Communication*, vol 15, no 4, pp. 469–500.

Wilson Rowe, E. (2009) 'Who is to blame? agency, causality, responsibility and the role of experts in Russian framings of global climate change', *Europe-Asia Studies*, vol 61, no 4, pp. 593–619.

Yagodin, D. (2010) 'Russia: listening to the wind – clientelism and climate change', in E. Eide, R. Kunelius and V. Kumpu (eds.), *Global Climate – Local Journalisms. A Transnational Study of How Media Make Sense of Climate Summits*, projekt verlag, Bochum, pp. 275–290.

Zhuravlev, I. and Leskov, S. (1998) 'Rossiia stanet krupneishim prodavtsom vozdukha', *Izvestiia*, 4 November.

6 Climate change in Russia's Far East

Controversial perspectives (mid 1990s–2010s)

Benjamin Beuerle

Introduction

During the last few years, a growing – though still not very big – number of scholarly texts have been published on Russia's relation to climate change. They cover such various topics as Russia's position in international climate change negotiations (Andonova and Alexieva, 2012), climate change in the Russian media (Poberezhskaya, 2015, 2016), political decision-making and debates on climate change (Andonova, 2008; Wilson Rowe, 2013; Korppoo, 2015, 2016), the history of Russian climate science (Oldfield, 2013, 2016) and, most recently, reasons for widespread climate scepticism in Russia (Kokorin, 2017; Tynkkynen and Tynkkynen, 2018). We know thus more and more – though still not enough – about the Russian position(s) on climate change at a national level. What we know so far about attitudes in the Russian public and among Russian scientists points to widespread climate scepticism (Dronin and Bychkova, 2017; Kokorin, 2017; Tynkkynen and Tynkkynen, 2018).[1] However, regional differences are until now markedly missing in this picture. It is a starting hypothesis of this chapter that attitudes towards climate change might differ depending on the situation in the various regions, their location and their respective affectedness by climate change. This concerns not least the Russian Far East. Its huge distance from the Russian capital; its relative closeness to countries like China, the Koreas, Japan, the United States and Canada; its various climates ranging from almost subtropical to (sub)arctic; its bordering on the Pacific Ocean; and its, at least partially, higher vulnerability to climate change (compared with Moscow and central Russia): these are all factors which could influence regional attitudes on the topic of climate change. The region is scarcely populated, but strategically important (Davis, 2003; Kotkin and Wolff, 1995; Blakkisrud and Wilson Rowe, 2018). It can be assumed that in this sense, Far Eastern attitudes could matter for Moscow's stance on climate change.

Against this background, it is the aim of this article to give some insights into positions on climate change in the Russian Far East and their development during the last two decades (mid-1990s–mid-2010s). What importance is given to climate change as a topic? Are there noticeable developments within the period of examination? Are there regional differences within the Russian Far East? Which effects of climate change on the Russian Far East have been discussed in the press? What is the mood today in the regional expert community, and to what extent can differences from the national level and Far Eastern particularities be discerned? In order to deal with these questions, two kinds of sources have been examined. The chapter is based above all on articles from the Far Eastern newspapers *Vladivostok, Zolotoi Rog* (both appearing in Vladivostok), *Kamchatskoe Vremia* (Petropavlovsk-Kamchatsky) and *Tikhookeanskaia Zvezda* (Khabarovsk) featuring the keywords 'climate change' and 'global warming'. These newspapers have been selected for their circulation (*Tikhookeanskaia Zvezda, Vladivostok* and *Kamchatskoe Vremia* have the biggest circulation for the respective regional centre town, at least from those available on *Integrum Worldwide*),[2] variety of ownership structure (*Tikhookeanskaia Zvezda* is owned partly by the staff, partly by the government of Khabarovsky Krai; *Zolotoi Rog* is a business-weekly held by a stock company; *Vladivostok* and *Kamchatskoe Vremia* are 'independent', privately owned and financed by ads) and availability.

The picture emerging from the analysis of these newspaper articles is developed and complemented by 18 interviews on climate change with scientists and with some leading activists of environmental NGOs conducted during a research trip to the Russian Far East in November 2017 in Vladivostok and Petropavlovsk-Kamchatsky. Among the 18 people interviewed, there were five activists from environmental NGOs and 13 professional scientists (among them 11 members of the Far Eastern Branch of the Russian Academy of Sciences (FEB RAS) and two staff members of state research facilities). Among the 13 scientists, there were four economists (all of them dealing in some way or the other with climate change) and nine natural scientists (two botanists, another biologist, two oceanographers, two hydro-meteorologists, one geographer, one geologist). This range and mix of professions can give insights into attitudes towards climate change on the part of the Far Eastern scientific and environmental communities as far as they are interested in climate change, but beyond the narrow set of climatologists primarily dealing with the issue.[3]

In terms of methodology, within the timespan available on *Integrum Worldwide* all articles from the abovementioned newspapers containing the keywords 'global warming' and 'climate change' have been consulted, with the exception of those where 'climate change' had other meanings (like a change in business climate or a pure change of weather circumstances). The timespan of availability was 1997–2015 for *Tikhookeanskaia Zvezda*, 1995–2015 for *Vladivostok*, 1996–2016 for *Zolotoi Rog* and 1999–2017 for *Kamchatskoe Vremia*. The interviews were conducted in an unstructured way, but all of them contained questions on the opinion of the interviewees about the topic of climate change and its (possible) origins, about its consequences and about possible regional particularities of the Russian Far East regarding this topic.

While the degree of interest in the topic within the regional newspapers indicates a strong influence of national – and partly international – factors, the content of the regional discourse on climate change is obviously much more influenced by specific regional factors. The results of the chapter point thus to regional – and interregional – particularities and differences within Russia which both warrant further investigation.

Coverage of climate change in Far Eastern newspapers, mid-1990s–mid-2010s

Why is it of interest to examine newspaper articles on climate change? The question must be asked, especially in a time when newspapers are arguably consulted only by a relatively small part of the population and the majority get their information from television and, recently, via the internet. Marianna Poberezhskaya has already given a number of pertinent answers to this question, while underlining also various problems with newspapers as primary research sources (Poberezhskaya, 2016: 13–64, 97–98). However, two more reasons should be given here for using these sources: 1) easy accessibility: this is a purely pragmatic reason, but for the purpose of scientific work, it is not without importance that newspaper articles are not only relatively easily and reliably accessible, but since the existence of databases like *Integrum Worldwide*, they can as well be searched through for keywords like 'climate change' and 'global warming' in a reasonable time. 2) Even more important is the multiplying factor of newspapers: newspaper articles can tell us much more than the opinions of the journalists. They cite, refer to (or directly include interviews) of scientists, politicians, public figures and societal activists etc. They can thus give insights into societal moods and opinions far beyond the position of the newspaper in question. This might be true to various degrees depending on the topic and the newspapers. It certainly applies in the case of climate change in the Russian Far East. Tellingly, as will be elaborated below, various and often controversial positions on climate change can be found in each of the four newspapers examined, independent of the regional and political orientation and owner-ship structure of these papers.

How and to what extent has the topic been covered in these media? It is a first finding that in all of the four newspapers examined, climate change (and/ or global warming) appeared in a number of articles and thus consti-tuted a discernible topic. I started the research with the hypothesis that it might be a bigger topic in the Far East than on a national level because of the region's greater vulnerability to climate change. This hypothesis is not confirmed by the data. Over a period of some 20 years, 118 articles men-tioning climate change appeared in Khabarovsk's *Tikhookeanskaia Zvezda* (1997 through 2015), 88 in *Vladivostok* (1995–2016), and 43 in *Zolotoi Rog* (1996–2016) and *Kamchatskoe Vremia* (1999–2017) each. That is, even in *Tikhookeanskaia Zvezda*, which had the widest coverage of climate change among the four newspapers and has the biggest circulation, there were on average only about six articles per annum. This is to be compared with some 400 articles each featuring climate change during the same time range in the big national newspapers *Kommersant* and *Izvestiia*. This difference can be explained by the fact that climate change is often considered as an international topic and that regional newspapers are usually more interested in regionally relevant topics.

A second finding concerns developments over time. Overall, in the 1990s there were only a very small number of articles mentioning the topic, whereas the number grew substantially after 2000. This corresponds with Poberezhs-kaya's findings for the national press (Poberezhskaya, 2015: 103–106). Inter-estingly, however, in all four regional newspapers examined, the number of corresponding articles reached a maximum somewhere between 2008 and 2012 and receded again afterwards. As this maximum varied between the four newspapers (2012 in *Tikhookeanskaia Zvezda*, 2009 in *Vladivostok*, 2008 in *Kamchatskoe Vremia*, 2009 in *Zolotoi Rog*), it can hardly be explained by some outstanding event like the 2010 heat wave or by the Copenhagen con-ference in 2009. If we consider Poberezhskaya's findings on the importance of national politics for the topics reported on by the press (Poberezhskaya, 2016: 38–58, 89), it is reasonable to assume that the abovementioned devel-opments are to be explained by the different agendas of Putin and Medvedev in their presidencies and by Medvedev's pushing of a Russian climate doc-trine in line with international efforts to combat climate change (2009) (ibid: 80–88). Structural reasons like the economic benefits of climate change miti-gation politics (Poberezhskaya, 2016: 83) would not be enough to explain the reverse trend from 2012. In other words: the personal agenda of the respective Russian president clearly impacts the attention given by the public to a topic like climate change, including on a regional level.

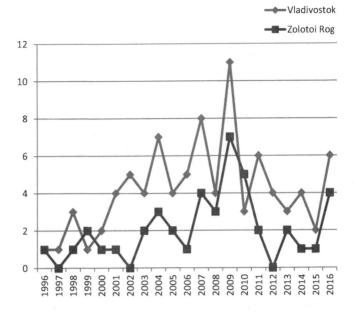

Figure 6.1a Number of articles referring to 'climate change' or 'global warming' over time

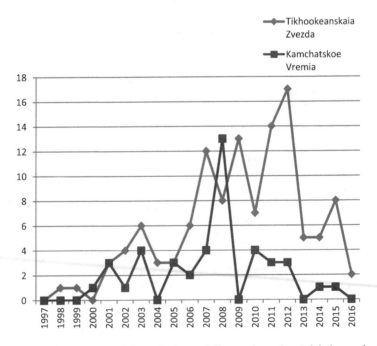

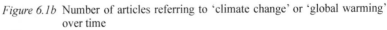

Figure 6.1b Number of articles referring to 'climate change' or 'global warming' over time

The regional dimension of climate change

Whereas the conjunctures of overall interest in climate change correlated with national politics, the content of the articles and more generally Far Eastern attitudes towards climate change have been in many respects influenced by regional concerns. In this sense, the following theses will be elaborated: 1. the Russian Far East is already affected by climate change; 2. apart from coverage of the topic in the regional press, engagement with climate change on the part of scientists, politicians and some companies in the region has at times been remarkably pronounced; 3. the region's position in the Asia-Pacific matters for this engagement with climate change; 4. climate scepticism in the region remains widespread – but not as dominant as recent literature suggests it is for Russia at large; and 5. as of late, interest in climate change is again on the rise at least in Primorsky Krai.

Impact of climate change on the Russian Far East

According to all accounts, climate change is proceeding faster in the Russian Far East than on world average or on Russian average (Vladivostok, 2004; Petrachkov, 2014). Connected with this fact, the region is already affected by climate change in various ways. To begin with, during the last 25 years, a growing number of extreme weather incidents have been ascribed to climate change by regional experts and in the press. This concerns hurricanes, heavy rainfalls and inundations, e.g. at the lake of Khanka in the eastern part of Primorie (e.g. Okovitaia, 2016; Zhuravlev and Klyshevskaia, 2016).

No less importantly, climate change has regularly been cited as a main reason for considerable changes of biodiversity in the region. This concerns above all the appearance of new fish species, which were traditionally to be found in more southern and warmer waters only. While several attacks on swimmers by white sharks (hitherto unseen in the region) in the summer of 2011 in the waters of Primorie seem to have been an exception until now – though an attention-grabbing one with disturbing short-term results for the tourism sector – (*Zolotoi Rog*, 2011a), the same cannot be said for a number of other species which are attracted by a rise of temperature in the Sea of Japan as well as in the waters around Kamchatka. At the same time, some of the fish species which were traditionally to be found in the region and are of importance for the regional fishing business have started to disappear or to be found in ever fewer numbers – a phenomena which is explained by experts in part directly by the changes in temperature and in part by the competition from new species from more southern waters (*Tikhookeanskaia Zvezda*, 2006; *Kamchatskoe Vremia*, 2010). On the other hand, there have also been reports on positive effects of climate change like an enforced fertility of agriculture and, on Kamchatka, new possibilities to grow e.g. apple and plum trees and other agricultural crops (*Kamchatskoe Vremia*, 2012; Tanas, 2012). In short, climate change is already affecting the region's ecosystem(s) and biodiversity, with economic consequences for those making a living from natural resources in one way or the other.[4]

Finally, melting of permafrost as an obvious consequence of (global) warming is of concern to the Russian Far East as well. While this is not the case for Primorsky and Khabarovsky Krai nor even for the parts of Kamchatka with a significant population (situated to the south of the peninsula), it is of concern to the huge Sakha (Yakutia) Republic which is situated on the north-western extremity of the Russian Far East. Here, a substantial part of the infrastructure is built on permafrost. The melting of permafrost is therefore putting in danger (and partly already damaging) housing, streets, factories and oil and gas pipelines. Additionally, as a result of this melting, huge amounts of methane are discharged from the soil into the air, which threatens in turn to vastly accelerate the greenhouse effect and thereby climate change (potentially creating a vicious circle) (*Vladivostok*, 2004; *Zolotoi Rog*, 2015).

All in all, the impact of climate change on the Russian Far East is already substantial. This suggests that interest in and engagement with climate change in the region might be relatively strong as well.

Engagement with climate change in the region

There are various regional actors whose engagement can be discerned here: the press and scientists as well as political and economic actors. As seen above, since the late 1990s, the topic has appeared in the regional press on a regular basis. The numbers of articles mentioning climate change and their development suggest that national and international factors play an important role in determining the extent of interest and coverage including in the regional press.

There has been a significant interest in climate change on the part of the regional scientific community – above all members of the FEB RAS – for years. The aforementioned changes in long-term climatic trends, biodiversity and the regional ecosystem have been incentives to deal with the topic, and a number of institutes of the FEB RAS dealing with questions of phenology and climate have got the necessary qualified (and interested) personnel and infrastructure to follow these incentives. This scientific interest and engagement found expression among others in regular conferences on the topic of climate change which were initiated by institutes of the FEB RAS as early as 1996 (Ignatenko, 2001).

Engagement with the topic in regional politics and the economy has been sporadic but at times remarkable. Thus, in March 2009, the mayor of Vladivostok Pushkarev, together with a number of youth organisations, and like other town heads in Russia called on the town's population to participate in the worldwide action 'Chas Zemli' ('Earth Hour') by switching off all electric devices for one hour at a given time in order to reduce greenhouse gas emissions and in order to point to the urgent problem of human-made climate change (Stanislavskii, 2009).[5] More significantly, already in November 2004 the governor of Khabarovsky Krai had ordered regional scientists of the FEB RAS to develop market-based mechanisms for reducing greenhouse gas emissions and to evaluate the output of such emissions by regional enterprises and cities (Savchenko, 2004).[6] This new engagement had to do with the newly ratified Kyoto Protocol. Some regional companies made active use of its provisions which allowed for projects reducing emissions to be traded internationally. Thus, in June 2005, Khaborovskenergo signed a treaty with the Danish Environmental Agency, which was to finance the conversion of one of the regional big power plants from coal to gas, resulting in an annual reduction of 1 million tonnes of CO_2. The regional press pointed out that it was the first such project to be realised in Russia and that prospects for other similar projects to follow in the Far East were good – a sign that at least temporarily, engagement with climate change – and its opportunities – in the Far East was stronger than in other Russian regions (Ilin, 2005). Early in 2012 it was reported that another heating plant, Khabarovsk-1, had already been completely readjusted from coal to gas. The report underlined that this modernisation was not only directly beneficial to the ecology of Khabarovsk, its surroundings and the Amur river, but that the according emission reductions could now be traded internationally as part of the first commitment period (2008–2012) of the Kyoto Protocol, allowing for substantial international investments in other energy- and emission-saving projects in the region (Ilin, 2012). Since Russia exited the second period of the Kyoto Protocol (from 2012) (Korppoo and Vatansever, 2012: 8), an important incentive and method for regional business to engage in climate change mitigation has ceased to exist.

In general, the cited newspaper articles show that part of the regional business – and science – communities are keen to seize on the opportunities which are offered by international climate politics and the topic of climate change at large. This kind of attitude is to be differentiated from those who see climate change itself as a blessing for the region. Some articles and experts stressed the positive aspects it would have, above all a higher fertility of agriculture and a reduction of heating costs (e.g. Alekseev, 2002). In recent years, a certain deputy of Vladivostok's town Duma has even advocated a project to artificially accelerate climate change in order to increase substantially the temperature in the region – and who is certain that the effects on agriculture, the tourist sector and the economy of Primorie in general would only be beneficial (Neshchedrin, 2014). It should be stated, however, that this plan is not taken seriously by any of the scientists to whom I spoke to in November 2017.

By contrast, the relatively widespread awareness of the importance of climate change as a topic and of the opportunities it offers are a lasting feature and have often been connected with the region's position in the Asia-Pacific.

The region's position in the Asia-Pacific matters

If interest in climate change and engagement with it in the Russian Far East have been at times remarkably pronounced, the region's location in the Asia-Pacific has doubtlessly had an influence in this interest in several respects. Firstly, the mentioned changes in biodiversity and their importance for the region are linked with its bordering on the Pacific, which has resulted in the fishing industry being traditionally an important branch of the regional economy (now one of the most vulnerable to climate change). Secondly, though the first big treaty in the framework of the Kyoto Protocol for reducing regional emissions was signed (in 2005) with a Danish agency, long before this, Japan had already influenced regional debates and politics in the same direction. Thus, already in October 1998, the regional press reported on a two-billion-dollar 'ecological credit' promised by the Japanese government for projects to reduce greenhouse gas emissions in Russia. According to the report, the Japanese government wished a substantial part of the credit to be spent in the Russian Far East (due to its closeness to Japan), where emissions from the coal-driven energy sector remained substantial (Drobysheva, 1998). When in November 2004, the governor of Khaborovsky Krai ordered an evaluation of the emissions from regional enterprises and cities, an article in *Tikhookeanskaia Zvezda* explained that while regional emissions remained substantial, the huge forests of Khabarovsky Krai were more than compensating for them and should be taken into consideration internationally. According to this article, Japan was so far the only country aware of its respective responsibilities and ready to invest 1–2 billion dollars in the modernisation of the Russian Far Eastern energy sector (Savchenko, 2004).

Additionally, throughout the period of investigation a number of political, scientific and environmental meetings and actions held in the region included participants of various Asian-Pacific states and had climate change on their agenda. This was the case for several APEC summits. As the regional press reported on these summits, they served as a reminder to the regional public that the topic of climate change was to be taken seriously (Avchenko, 2007). PICES, the North Pacific Marine Science Organization, which was established in 1992 and has got Russia, Japan, China, South Korea, the United States and Canada as member states, has also regularly dealt with climate change and its consequences for the North Pacific Region in the form of working groups and international conferences, which were reported on in the regional press (Nazarov and Snytko, 2005; Zolotoi Rog, 2011b). Moreover, regular ecological forums, held annually in Vladivostok from 2006, and some actions on the part of environmental activists in the region each have integrated participants from Japan, South Korea and other Asia-Pacific countries and had the topic of climate change on their agenda. This underlines both the engagement with the topic of climate change on the part of various actors in the region and the importance of the Asia-Pacific context for the attention which has been given to the topic in the Far East by part of the public, scientific community and politico-economic actors (Lukonina, 2006; *Vladivostok*, 2007; *Zolotoi Rog*, 2016).

The (limited) importance of climate scepticism in the region

While a number of actors on various levels in the region have been interested in the topic of climate change and seem to be convinced of its relevance including for the region itself, climate scepticism remains widespread among the regional media and expert communities. Two different kinds of climate scepticism can be discerned here. First, there is a position doubting that global warming is going on. This position could be found from time to time in newspaper articles notably during the first years of the new millennium. Several articles cited experts who were of the opinion that we were rather living in a period of global cooling than of global warming (Ivleva, 2003; Grigor'evna, 2005). As late as October 2012, an article in *Tikhookeanskaia Zvezda* underlined that it was 'a matter of dispute' among scientists which view was correct (Savchenko, 2012). On his part, Boris Kubai, the head of Primorie's agency for hydrometeorology and observation of the environment, had reassured readers of *Vladivostok* during the heat wave of 2010 that there was no global warming, since overall, the planet was always in balance: if it was especially warm in this region, it was especially cold in others (Nadeina, 2010). Altogether, this kind of position could be found only in a relatively small number of articles (in 33 out of a total of 292 articles examined), but it reappeared sporadically including in recent years (Kochugov, 2016).

More common is the second version of climate scepticism, a position doubting that climate change was primarily anthropogenic. This was not evident from the articles examined, where this position found expression in less than 8% of the total. In each of the four newspapers, throughout the period of examination, there were more articles mentioning the anthropogenic nature of climate change.

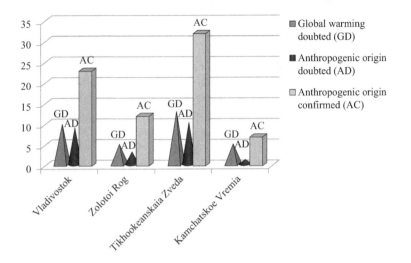

Figure 6.2 Overall number of articles doubting/confirming global warming or its anthropogenic origin (mid-1990s-mid-2010)[7]

The majority of articles refrained from taking any position on this question and just reported about climate change and its consequences without reflecting on its causes. How widespread this second climate sceptic position still is, became thus clear only through the interviews conducted in November 2017 in the Russian Far East. Among 18 scientists and environmental activists whom I interviewed in Vladivostok and in Petropavlovsk-Kamchatsky, eight doubted explicitly that climate change was human-made, and only seven stated their support for 'Al Gore's Hypothesis' – as the link between anthropogenic emissions and climate change was called by several interviewees. The main arguments brought forward by those espousing this kind of sceptics were the same in the interviews and in the newspaper articles. They also coincided in part with those offered by Dronin's and Bychkova's Russian interviewees (Dronin and Bychkova, 2017): (1) scientific observations had started only about hundred years ago. Decades or even hundreds of additional years of observation were necessary in order to come to reliable conclusions about the nature of climate change. (2) There had been substantial climatic changes in pre-industrial times. In most cases, this argument was connected with a circular (hypo) thesis: periods of warming and cooling were coming and going in circular rhythms, and there was nothing human beings could do about it. One of the interviewees claimed that global warming had been going on from the

1970s until around 2000 but had stopped since then – another sign for the disconnectedness between anthropogenic emissions and climate change. (3) Natural factors, like an alteration in the sun's activity or in the positions of the magnetic poles, were as good an explanation. (4) Various natural sources of methane gas (e.g. between tectonic plates) emitted much more than was caused by human beings. (5) The topic was artificially pushed and politicised by (Western) countries and experts whose financial interests or career prospects were directly linked to it. Several of my interviewees (all of them scientists) added another argument: (6) Donald Trump also doubted anthropogenic climate change, and the US President had certainly competent advisers.

How to make sense of the wide propagation of this kind of climate scepticism among regional experts? To begin with, the reasons offered by Nikolai Dronin and Alina Bychkova are pertinent in this case as well, particularly their thesis that the empiricist rather than rationalist orientation of many Russian naturalists leads to their distrust of findings which are not founded on direct observation (Dronin and Bychkova, 2017). One other main reason is presumably that the interviewed experts are professionally dealing with the consequences of climate change. They are – mostly – phenologists, not climatologists. They are thus interested in the phenomenon of climate change, but their opinions about the reasons for climate change are based on hypotheses rather than on theses verified scientifically by themselves. This is where the widespread climate sceptic discourse in Russia, as analysed by Nina and Veli-Pekka Tynkkynen (2018) as well as by Alexey Kokorin (2017), comes in. Reasons named by these authors include the active promotion of this discourse by lobby groups (private or state-based) connected with the fossil industry which is still vital for Russia's economy, as well as mistrust in an international mainstream position on climate change ('Al Gor'es Hypothesis') which seems to be very much fostered by the 'West' (Kokorin, 2017; Tynkkynen and Tynkkynen, 2018). Remarkably enough, the abovementioned argument that Trump has doubted anthropogenic climate change is at least partly in contradiction of the claim that global warming is a political idea designed to advance Western interests. Or else: Donald Trump is seen (arguably reasonably so) as a Western leader outside the Western mainstream.

Against the background of Tynkkynen's and Kokorin's analysis of widespread scepticism in Russia (and especially against the findings of Dronin and Bychkova regarding the dominance of climate scepticism among Russian environmental scientists), it is however even more noteworthy that climate scepticism is neither really dominant among the Far Eastern expert community (let alone among the newspaper articles analysed) nor is it coincident with the opinion that there is no need to reduce anthropogenic (greenhouse gas) emissions. On Kamchatka, five of the eight interviewees turned out to be explicit climate sceptics (among them both environmental activists with whom I talked to on the Peninsula), but in Vladivostok, this was the case for no more than four of the ten interviewed experts, with five interviewees making explicit their support for 'Al Gore's Hypothesis'. Moreover, two of the climate sceptics in Vladivostok underlined that while in their opinion it was reasonable to doubt the anthropogenic origins of climate change, they were nevertheless supportive of measures for reducing greenhouse gas emissions – just in case and because in most instances these emissions were harmful for human beings and the environment anyway. There is a case for regional differences within Russia and for interregional differences within the Russian Far East here. Admittedly, the sample and database used for this chapter is too small to offer more than preliminary findings, but so far they point to the conclusion that while Kamchatka fits into the overall Russian picture as described in recent literature, Vladivostok and Primorsky Krai do not. This corresponds with the fact that of late interest in climate change in the scientific community of Primorie is clearly on the rise.

Renewed interest in climate change in parts of the Russian Far East

In Vladivostok at least, there has recently been growing attention to the topic of climate change both on the part of the local scientist-community and on the part of some decision-makers. In October 2017, there was a big international conference on various aspects of climate change in Vladivostok,[8] and the organisers of this conference indicated to me that similar conferences will be held every second year from now on. An important role in this respect is played by the Italian Nobel prize winner Riccardo Valentini. In recent times, he has been regularly in Vladivostok and has initiated the establishment of a 'Climate Smart Lab' at Vladivostok's Far Eastern University with the declared goal of creating the leading climate change science-institution within the Asia-Pacific. When signing the contract for this establishment, Valentini underlined the crucial importance of doing research for mitigating – and adapting to – climate change in a balanced way, while the Far Eastern University's director Sergey Ivanets stressed above all the strategic potential of the new lab for enhancing the university's national and international standing (PrimaMedia, 2016a, b). Climate change has thus been identified not only as an important topic, but also as a strategically promising one. Including at the School of Economy and Management of the abovementioned university, a number of initiatives in the research and study of topics related to (anthropogenic) climate change have begun and are still being developed. This indicates that the renewed interest in climate change, the international orientation of climate change engagement in the region, and the opportunities which are offered by the topic of climate change to the regional science and potentially business communities can contribute to reducing the influence of climate sceptic discourse and to enhance the influence of 'Al Gore's Hypothesis' in Far Eastern dealings with climate change.

Concluding remarks

This article has provided a preliminary look at the topic of climate change awareness in the Russian Far East. According to Nina and Veli-Pekka Tynkkynen's (2018) as well as Nikolai Dronin's and Alina Bychkova's (2017) assessment, climate scepticism remains dominant in the Russian public and among Russian environmental scientists. However, whereas Kamchatsky Krai fits into this picture, Vladivostok does not. The region's location in the Asia-Pacific in combination with external and extraordinary factors like the influence and prestige of Riccardo Valentini and the frequent discussion of climate change during APEC and other international meetings held in the region play an important role here. Additionally, internal factors like the rich biodiversity of Primorie and the active research community which studies it and grapples with elements that change and influence it, a relatively mild climate as well as a growing number of extreme weather appearances are all of importance also. Most importantly, climate change is already discernibly affecting the Russian Far East, and at the same time climate change politics and the international attention to the topic offer opportunities for the regional science and business communities alike. Both factors have acted as incentives for taking the topic seriously. There are thus indicators that at least in Primorsky Krai specific regional factors might outweigh the 'hydrocarbon' culture which Veli-Pekka Tynkkynen has identified for Russia at large (see Chapter 4).

More research is certainly warranted to develop our understanding. It remains to be seen to what extent the abovementioned findings and trends can be confirmed by further research in the region, which of the centretowns under examination is more representative of the Russian Far East, and whether Primorie is, in this respect, a trendsetter for Russia, or whether it remains an ephemeral exception. One point has become clear though. If we want to understand Russian attitudes and approaches to climate change, it is worthwhile to take a closer look at regional peculiarities and differences.

Notes

1 See as well Veli-Pekka Tynkkynen´s chapter in this volume.
2 *Integrum Worldwide* is a continuously updated database integrating most Russian regional and national newspapers from the mid-1990s until the most recent times.

3 I started the interviews initially without a (manifested) intention to publish their results, which is one of the main reasons why I decided to anonymise them.
4 This concerns not least the indigenous peoples in the region – a topic which will have to be treated in another article (for literature on the topic, see Sharakhmatova, 2011; Graybill, 2013).
5 This action was to be repeated in similar ways in later years (Vladivostok, 2014).
6 I have still to investigate what followed from this initiative.
7 Including indirect confirmations (e.g. an article on the Copenhagen Conference and their goals would be counted as 'AC' including if the author of the article didn't confirm explicitly that these goals were justified).
8 'Climate Change Constraints and Opportunities in the Asia-Pacific Region: Human-Biosphere-Atmosphere Interactions and Green Growth': Conference held 24–26 October 2017 in Vladivostok (Far Eastern Federal University and Botanical Institute of the Far Eastern Branch of the Academy of Sciences).

References

Alekseev, A. (2002) 'Normal'no-anomal'naia zima', *Vladivostok*, 6 March.
Andonova, L.B. (2008) 'The climate regime and domestic politics: the case of Russia', *Cambridge Review of International Affairs*, vol 21, no 4, pp. 483–504.
Andonova, L.B. and Alexieva, A. (2012) 'Continuity and change in Russia's climate negotiations position and strategy', *Climate Policy*, vol 12, no 5, pp. 614–629.
Avchenko, V. (2007) 'Prinimaem estafetu Sidneia', *Vladivostok*, 11 September.
Blakkisrud, H. and Wilson Rowe, E. (eds). (2018) *Russia's Turn to the East. Domestic Policymaking and Regional Cooperation*, Palgrave Macmillan, Cham.
Davis, S. (2003) *The Russian Far East. The Last Frontier?* Routledge, London.
Drobysheva, I. (1998) 'Dostanetsia li Primor'iu Iaponskii kredit?' *Zolotoi Rog*, 31 October.
Dronin, N. and Bychkova, A. (2017) 'Perceptions of American and environmental scientists of today's key environmental issues: a comparative analysis', *Environment, Development and Sustainability*, https://doi.org/10.1007/s10668-017-9979-8.
Graybill, J.K. (2013) 'Imagining resilience: situating perceptions and emotions about climate change on Kamchatka, Russia', *GeoJournal*, vol 78, no 5, pp. 817–832.
Grigor'evna, T. (2005) 'Stsenarii dlia pogody', *Vladivostok*, 18 August.
Ignatenko, I. (2001) 'Teplo i syro', *Vladivostok*, 24 August.
Ilin, V. (2005) 'Amurskaia TETs-1: Den'gi iz vozdukha', *Tikhookeanskaia Zvezda*, 30 June.
Ilin, V. (2012) 'Zelennye investitsii pridut v Khabarovsk', *Tikhookeanskaia Zvezda*, 4 February.
Ivleva, M. (2003) 'Bunt zemli', *Vladivostok*, 14 March.
Kamchatskoe Vremia (2010) 'Solnechnaia ryba', *Kamchatskoe Vremia*, 23 December.
Kamchatskoe Vremia (2012) 'I na Kamchatke rastut duby', *Kamchatskoe Vremia*, 24 October.
Kochugov, V. (2016) 'Gorod v zolotoi seredine', *Vladivostok*, 8 June.

Kokorin, A. (2017) 'Analiz problemy klimaticheskogo skeptitsizma', *Ispol'zovanie i okhrana prirodnykh resursov v Rossii*, no 4, www.priroda.ru/upload/iblock/01a/3 статья.doc.

Korppoo, A. (2015) 'Russia's climate policy', in G. Bang, A. Underdal and S. Andresen (eds.), *The Domestic Politics of Global Climate Change*. Edward Elgar Publishing, Cheltenham, pp. 141–159.

Korppoo, A. (2016) 'Who is driving Russian climate policy? applying and adjusting veto players theory to a non-democracy', *International Environmental Agreements: Politics, Law and Economics*, vol 16, no 5, pp. 639–653.

Korppoo, A. and Vatansever, A. (2012) 'A climate vision for Russia. From rhetoric to action', *Carnegie Policy Outlook*, August, http://carnegieendowment.org/files/RussiaClimate.pdf

Kotkin, St. and Wolff, D. (ed). (1995) *Rediscovering Russia in Asia: Siberia and the Russian Far East*. Taylor and Francis, Hoboken.

Lukonina, E. (2006) 'Primorskaia rossyp´ prirodnykh zhemchuzhin', *Vladivostok*, 9 June.

Nadeina, T. (2010) 'Anomal'noe leto', *Vladivostok*, 25 August.

Nazarov, V. and Snytko, V. (2005) 'PICES – nadezhda na spasenie okeana', *Vladivostok*, 1 November.

Neshchedrin, M. (2014) 'Kogda k nam prishli leopardy', *Vladivostok*, 20 August.

Okovitaia, N. (2016) 'More progreetsia raz . . .', *Zolotoi Rog*, 16 August.

Oldfield, J.D. (2013) 'Climate modification and climate change debates amongst Soviet physical geographers, 1940s – 1960s', *WIREs Climate Change*, vol 4, no 6, pp. 513–524.

Oldfield, J.D. (2016) 'Mikhail Budyko's (1920–2001) contributions to global climate science: from heat balances to global climate change and global ecology', *WIREs Climate Change*, vol 7, pp. 682–692.

Petrachkov, S. (2014) 'Ne grozit li Iaponskomu moriu sud'ba Chernogo?' *Vladivostok*, 8 October.

Poberezhskaya, M. (2015) 'Media coverage of climate change in Russia: governmental bias and climate silence', *Public Understanding of Science*, vol 24, no 1, pp. 96–111, http://journals.sagepub.com/doi/pdf/10.1177/0963662513517848.

Poberezhskaya, M. (2016) *Communicating Climate Change in Russia. State and Propaganda*. Routledge, London.

PrimaMedia. (2016a) 'Nobelevskii laureat vozglavil laboratoriiu issledovaniia klimata v DVFU', *Primamedia.ru*, 1 March, https://primamedia.ru/news/492540/.

PrimaMedia. (2016b) 'Rikkardo Valentini: "Parnikovye gazy groziat naseleniiu Zemli global'nym golodom" ', *Primamedia.ru*, 14 March, https://primamedia.ru/news/494520/.

Savchenko, A. (2004) 'Nash les ne dast ves' mir v parnuiu prevratit' ', *Tikhookeanskaia Zvezda*, 17 November.

Savchenko, A. (2012) 'Prichinu pozharov ishchi v okeane', *Tikhookeanskaia Zvezda*, 10 October.

Sharakhmatova, V. (2011) *Nabliudeniia Korennykh Narodov Severa Kamchatki za Izmeneniiami Klimata. Otchet*. Kamchatpress, Petropavlovsk-Kamchatskii.

Stanislavskii, A. (2009) 'Poka gorit svecha . . .', *Vladivostok*, 31 March.

Tanas, O. (2012) 'Dal'nii Vostok sdaetsia v arendu', *Tikhookeanskaia Zvezda*, 31 January.

Tikhookeanskaia Zvezda. (2006) 'Pochemu ryba v osnovnom v brakon'erskie seti plyvet?' *Tikhookeanskaia Zvezda*, 30 November.

Tynkkynen, V.-P. and Tynkkynen, N. (2018) 'Climate denial revisited: (Re)contextualising Russian public discourse on climate change during Putin 2.0', *Europe-Asia Studies*, doi:10.1080/09668136.2018.1472218.

Vladivostok. (2004) 'Goriachie tochki', *Vladivostok*, 8 April.

Vladivostok. (2007) 'Ochistiat pliazh i ostanoviat poteplenie', *Vladivostok*, 25 July.

Vladivostok. (2014) 'Chas bez sveta za ekologiiu', *Vladivostok*, 1 April.

Wilson Rowe, E. (2013) *Russian Climate Politics: When Science Meets Policy*. Palgrave Macmillan, Basingstoke/New York.

Zhuravlev, Iu. N. and Klyshevskaia, S. V. (ed). (2016) *Transgranichnoe ozero Chanka: prichiny povysheniia urovnia vody i ekologicheskie ugrozy. Materialy I Dal'nevostochnoi konferentsii, 27–29 aprelia 2016g.*, Dal'nauka, Vladivostok.

Zolotoi Rog. (2011a) 'Akuly isportili sezon v Primorie', *Zolotoi rog*, 23 August.

Zolotoi Rog. (2011b) 'Pogoda na Tikhom Okeane uvelichit [. . .]', *Zolotoi Rog*, 8 November.

Zolotoi Rog. (2015) 'Zhara pridet iz . . . Sibiri', *Zolotoi Rog*, 21 July.

Zolotoi Rog. (2016) '"Priroda bez granits": desiat' let v formate dialoga', *Zolotoi Rog*, 25 October.

7 Russian industry discourses on climate change

Ellie Martus

Introduction

Industry and business actors are key players in the fight against climate change. They hold significant responsibility when it comes to greenhouse gas (GHG) emissions, and their participation is vital in global efforts to reduce emissions. They can be powerful political and economic actors, with the ability to shape government policy and wider discussions in society on climate. It matters, therefore, what these companies say and do in relation to climate change.

This discussion will concentrate specifically on oil and gas companies. Overall, the energy sector is the largest contributor to Russia's GHG emissions. According to the UN Framework Convention on Climate Change (UNFCCC), the energy sector was responsible for 82.16% of Russia's total GHG emissions (without land use, land use change and forestry) in 2012 (UNFCCC, 2015). The key question this chapter seeks to explore relates to how Russian industry understands the issue of climate change. That is, do companies acknowledge climate change as an issue? If so, how do they talk about climate change? How have they understood and conceptualised the issue? What are their priorities, how do they seek to address climate change, and what policy solutions do they advocate? It is about the disclosure of information, corporate understanding and framing, and industry narratives on climate change. Understanding industry narratives on climate change is an important element of Russia's broader climate change discourse.

There is a substantial body of literature which explores the relationship between climate, politics and business. The role of business in governing climate change and shaping policy has featured heavily, with works examining the push for self-regulation and other forms of private governance (Hoffman, 2005; Bulkeley and Newell, 2015), support for emissions trading schemes (Meckling, 2011) and the emphasis on management processes, policy influence and image rather than meaningful GHG reductions (Jones and Levy, 2007), for example.

As early entrants into global climate politics, fossil fuel companies have been the focus for much research, with works charting the aggressive lobbying against action on climate change by companies such as Exxon and Chevron (Newell and Paterson, 1998: 683). After this initial opposition, attention shifted to the emergence of a second stream of corporate political involvement, with Kolk and Pinkse noting the move towards engagement with market responses to climate change (2007: 202). These divergent policy responses from the oil industry have been charted by Pulver (2007). Research has also expanded on different strategies adopted by oil multinationals (Levy and Kolk, 2002; van den Hove et al., 2002; Levy, 2005).

While Russia has been largely excluded from this literature, a limited number of works have touched on the role of Russian business actors in climate politics, including through their support for ratification of the Kyoto Protocol (Andonova, 2008), and interest in Joint Implementation projects (Korppoo, 2007). The attitudes of specific industrial sectors within Russia towards climate change has been thus far overlooked, with the exception of work by the present author on the metals and mining sector which pointed to the strong resistance to climate policy from the coal industry (Martus, 2018).

This chapter also draws on literature which explores discourse on climate change, particularly from a business or industry perspective. Works have explored a variety of themes, including for example, ExxonMobil's corporate discourse on global warming, with Livesey examining how public advertisements were used by the company to 'promote particular corporate understandings of the problem of the natural environment and legitimate the corporate stance' (2002: 118). There are a number of competing frames, or discourses, on climate change: including Hajer's (1995) 'environmental modernization'. According to this, climate change action is framed within an economic development paradigm (Taylor, 2013: 26). As Levy notes, this discourse 'puts its faith in the technological, organizational, and financial resources of the private sector, voluntary partnerships between government agencies and business, flexible market-based measures, and the application of environmental management techniques' (2005: 93). Climate change becomes a problem that can be solved through the market. This chapter explores how the Russian experience fits within this broader context.

This chapter proceeds as follows. A brief introduction to the core tenets of Russian government policy and relations with the oil and gas sector is given. Second, methodological considerations arising from this study are discussed. An analysis of climate change discourse within the oil and gas sector is then presented, followed by a discussion of key implications arising from this research, and conclusions.

Oil, gas and the state in Russia

The oil and gas sector plays a central role in Russia's economy. In 2016 for example, oil and gas revenues brought in 4 844 billion roubles to the federal budget, or 36% of total revenue for the year (Ministerstvo Finansov, 2018). There are a number of large companies operating in the sector, several of which are examined below. These companies are both private and state-owned, with the oil and gas sector having undergone considerable changes since the end of the Soviet Union. The sale of previously state-owned enterprises in the 1990s (including the notorious 'loan-for-shares' auctions) saw the oil sector largely privatised by the end of 1998 and fragmented (Gustafson, 2012: 98–99). The industry then began a process of reconsolidation, with private oligarch-owned companies like Lukoil, Surgutneftegaz and Yukos emerging (Gustafson, 2012: 100–101). Since 2002, however, there has been an increasing state presence in the industry, with Rosneft emerging as a particularly powerful actor. The gas sector in Russia is even more concentrated and dominated by state-owned company Gazprom.[1]

While Russia's overall approach to climate policy has been described by other authors in this volume, there are a number of policies of direct relevance to the oil and gas sector. Some of these policies directly address climate change mitigation, while others touch on related issues such as energy efficiency and the development of renewables. Ultimately, however, the primary goal of Russian energy policy is to actively support the industry. The principle policy document pertaining to the oil and gas sector is the *Energy Strategy to 2030*. The stated primary goal is to maximise the efficient use of energy resources to ensure economic growth. Reference is made in the document to the need to reduce GHG emissions in the energy sector. Specific legislation that obliges companies to report their GHG emissions is currently under development (Davydova, 2017; Pravitel'stvo Rossii, 2018).

Russia's energy sector is highly inefficient by global standards, and consequently, there have been a number of attempts to introduce energy saving and efficiency measures. One of the most important of these is a decree (*Ukaz 889*) issued by then President Medvedev on 4 June 2008 titled 'On some measures to improve the energy and environmental efficiency of the Russian economy'. This aimed to reduce the energy intensity of the economy by at least 40% by 2020 and called on the government to introduce a number of measures including new laws to encourage businesses to use energy-saving technology. Beyond this, the government introduced regulations in 2012 on the use of associated petroleum gas (APG), which has often been considered a waste product by the oil and gas industry and burnt. A number of regulations have been put in place by the government to limit this practice, including a 2012 decree restricting flaring to no more than 5% (Korppoo and Kokorin, 2017: 120–121). There have also been some attempts to encourage the use of and investment in renewable energy, with a number of government orders and decrees introduced (see Korppoo and Kokorin, 2017: 121). Despite considerable potential, however, growth in renewables has been slow (IEA, 2014: 219). The analysis that follows considers the extent to which these policies have had an effect on Russia's oil and gas companies in shaping their understanding and response to the issue.

Methodology

The Russian oil and gas sector has been chosen as the focus for analysis, given its significance for the Russian economy, and the sector's role as a major producer of GHG emissions. There is an additional goal of collating, comparing and analysing reporting from major Russian oil and gas companies and making the contents of these reports available to a wider audience.

To evaluate corporate responses to climate policy, a survey of the eight largest oil and gas companies according to the Ekspert 400 rankings (Ekspert.ru, 2017) was undertaken (see Table 7.1 below). Together, these companies represent a substantial portion of Russia's oil and gas sector. Gazprom alone accounts for 66% of Russian gas output, which is 11% of global production (Gazprom, n.d.), and Rosneft is responsible for 40% of Russia's oil production, and 6% of global production (Rosneft, n.d.). Companies are both private, such as Lukoil, and state-owned corporations, such as Gazprom and Rosneft.[2]

To assess what information about climate change mitigation efforts is disclosed, and identify corporate discourses on climate change, analysis was based on Sustainability Reports (SR), Environment Reports (ER) or Annual Reports (AR) where SRs or ERs were not available. Company websites

Table 7.1 Russia's largest oil and gas companies

Company	Volume of Sales in 2016 (million rubles[3])
Gazprom XE "Gazprom"	6 071 793,0
Lukoil	4 743 732,0
Rosneft	4 134 000,0
Surgutneftegaz	1 020 833,0
Tatneft	580 127,0
Novatek	537 472,0
Bashneft	494 722,0
Sakhalin Energy	304 810,0

(Source: Ekspert.ru, 2017)

were also examined. Sustainability, environmental and annual reports are generally prepared in both Russian and English (with the exception of Bashneft, where only a Russian language SR was available). Where the English version has been used in this chapter, it has been checked against the Russian for consistency. Websites too are, for the most part, in both Russian and English. Some companies also prove other languages too: Sakhalin Energy, for example, has a Japanese-language version.[4]

These reports, together with company websites, represent key forms of corporate communication. They enable corporations to advertise themselves and their agenda. This advertising is, for the most part, directed at company stakeholders, including investors, shareholders and the government, but also to the wider community. There is a vast literature on Corporate Social Responsibility (CSR) reporting, much of which concentrates on identifying motivations for firms in issuing these voluntary documents. This might be, for example, to demonstrate their commitment to the principles of CSR, or for 'greenwashing' purposes to influence stakeholder perceptions (Mahoney et al., 2013). There is also considerable discussion about the credibility of information (e.g. Michelon et al., 2015). The lack of independent verification of reported results and activities in these reports presents something of a challenge. However, despite issues with determining the accuracy of information, and the scepticism surrounding motivations for CSR disclosure, they remain a valuable tool for corporate actors to demonstrate their commitment to CSR (Allen and Craig, 2016). For the purposes of this research, corporate reporting on the environment from the Russian oil and gas sector helps us assess how companies want to be perceived by external actors and how the companies themselves regard certain issues. That is, are Russia's oil and gas companies interested in generating positive environmental PR by demonstrating how they address climate change, or is it low on the list of priorities?

The majority of companies provide SRs. These reports mostly follow a similar format, based on international reporting standards.[5] These documents provide information on corporate activities, including production results for the year, an overview of assets, new projects and future developments, and corporate governance. They also discuss environmental impact and management issues of the company, outline environmental priorities and spending. Reports cover a wide range of issues, including air, water and soil protection, waste management and biodiversity conservation. Projects involving community groups and non-governmental organisations (NGOs) are discussed. Personnel, and health and safety issues are also covered in the SR and AR documents. Some companies, such as Gazprom and Surgutneftegaz, provide stand-alone environment reports, which concentrate specifically on company environmental policies and activities.

Based on an analysis of the available documents, a series of key themes emerged that touched on the issue of climate in some form or another. Both climate specific (such as direct statements or positions on climate change, or reporting on GHG emissions and mitigation efforts) and climate-related material (such as energy saving and efficiency measures, and investment in renewables) were included. Information about corporate non-climate environmental programmes was also included for comparative purposes. These themes highlight the priorities of companies and their understanding of the issue and are used to build a picture of industry discourse on climate change.

Industry framings of climate change

Examining the environmental reporting contained within AR, SR and ER from Russia's oil and gas majors provides us with a valuable insight into how these large companies conceptualise the issue of climate change, and how companies think the issue should be addressed. This section first discusses direct references to climate change and identifies those corporate actors with clear positions on climate. Second, GHG emission reduction efforts are discussed together with the related issue of APG utilisation. Finally, corporate energy-saving and efficiency measures are discussed.

Stand-alone position on climate

Of the companies surveyed, only Lukoil had a stand-alone position on climate change, although even this is limited in terms of what it covers. Lukoil's 'stance on climate change' states that the company takes an active part in discussions with the government on GHG regulation, and notes that Lukoil has developed its own system for managing emissions to respond to the issue of climate change (Lukoil, n.d. -c). A number of GHG reduction projects initiated by Lukoil are listed. Framed in market-based terms, it is noted that if these projects are successful, the 'ultimate goal could be to generate revenues by selling emission reduction units and to further reinvest the generated funds into new environmental and energy efficiency projects' (Lukoil, n.d. -c).

Further to its official 'stance' on climate, on the English version of the company's website, Lukoil 'as an environmentally responsible company, PJSC LUKOIL admits that global climate change prevention effort is extremely important' (Lukoil, n.d. -b). Similarly, on the Russian version of the website the company 'recognises the importance of measures to prevent global climate change, supports Russia's participation in global efforts to reduce emissions of greenhouse gases' (Lukoil, n.d. -c). Similar statements exist in the 2015–6 SR, where it is noted, for example, that 'LUKOIL acknowledges the importance of combating global climate change, and supports Russia's contribution to the global effort to reduce greenhouse gas emissions' (Lukoil, 2015–6: 49). It is significant that Lukoil establishes its support for international climate policy efforts and Russia's participation, as it is the only oil and gas company examined to explicitly do so.

However, despite this attention given to expressing direct support for climate policy efforts, Lukoil's strategies to address the issue are very similar to the other companies examined. For example, Lukoil's *Health, Safety and Environmental Policy* includes climate-related goals including achieving a 'higher utilization ratio of associated petroleum gas', the 'increased output of eco-friendly fuels compliant with European standards', 'efficient control of greenhouse gas emissions' and the efficient use of natural resources through the 'introduction of resource-saving and energy efficient technologies and use of alternative energy sources' (Lukoil, n.d. -a). Other oil and gas companies also address these issues, as we will explore below.

While lacking a stand-alone company position on climate, a number of the other companies include statements addressing climate change as an issue in their corporate reporting. The deputy chair of Gazprom's Management Committee, Vitaly Markelov, for example, stated that 'Gazprom constantly improves the energy efficiency and reduces anthropogenic impact on the environment through the application of advanced technologies and equipment' and that the company 'takes part in solution [*sic*] of global air pollution issues by promoting natural gas as an environmentally safe motor fuel. Owing to the use of natural gas, energy balance of Russia is one of the "greenest" in the world' (Gazprom, 2016: 5). Again, we see a focus on energy efficiency as a central policy solution, and an emphasis on new technology. Markelov's statement also introduces the idea of natural gas as a green, climate-friendly source of energy, a sentiment (unsurprisingly) echoed by the big gas producers. Novatek too, for example, states that an intended aim of its environmental policy is to develop the market for 'environmentally safe' gas-engine fuels (Novatek, 2016b).

As we might expect, the need to balance climate mitigation efforts with economic concerns is a core element of the industry narrative. This is highlighted by Novatek, for example, which points to company participation in the Carbon Disclosure Project (CDP) and disclosure of information on GHG emissions and energy efficiency. Novatek's AR states that 'by taking part in these projects the Company intends to achieve a balance between the climate change risks and efficiency of investment projects' (Novatek, 2016a: 55). Surgutneftegaz's report also briefly mentions climate change, noting simply that 'the Company implements measures on prevention of global climate change, develops the corporate system of accounting and management of greenhouse gas emissions' (2016: 34).

Interestingly, there was little evidence to suggest climate change was considered a major risk to business operations to any large extent by the companies in question, despite the challenges posed by a changing climate, particularly in terms of physical risk to infrastructure and equipment. This is despite the frequent use of discourse on 'risk' to frame corporate responses to climate change observed by Wright and Nyberg for example (2015). The closest any company came to this idea of risk was Novatek, in its Environmental Policy. The policy sets out a number of environmental objectives and aims to 'take into account the risks and assess the consequences of climatic changes for the activities of the company' (Novatek, 2016b). However, all companies did address physical risks in broader terms (oil spill and emergency situation preparations for example) in the ARs & SRs.

Finally, the companies examined did not appear to be heavily engaged with international developments in climate policy. Three companies (Gazprom, Bashneft and Sakhalin Energy) made reference to the Paris Agreement, but there was no discussion of content or implications for company operations (Bashneft, 2015: 44; Gazprom, 2016: 68; Sakhalin Energy, 2016: 72). Some clues were provided regarding company involvement in international bodies, with Rosneft, Lukoil and Sakhalin Energy, referring to participation in the UN Global Compact, a voluntary initiative designed at promoting sustainable business practices. They are the only Russian oil and gas companies involved, although none participate in the 'Caring for Climate Initiative' which is specifically aimed at addressing corporate responses to climate change.

GHG emissions and reduction efforts

All companies surveyed reported their GHG emissions or air pollutant emissions (a broader category). Of note, however, is the fact that there were inconsistencies between the companies surveyed in regards to what is reported and how emissions are calculated. Sakhalin Energy, for example, calculates its emissions in accordance with the guidelines developed by the American Petroleum Institute (Sakhalin Energy, 2016: 72), while Novatek and Bashneft reports that they use the 2015 Russian government guidelines (Bashneft, 2015: 45; Novatek, 2016c: 76;). Four companies also report their emissions to the CDP: Lukoil, Gazprom, Surgutneftegaz and Novatek.

The lack of consistency in relation to GHG reporting is due to a gap in government policy. In January 2018, the government's commission on legislative activity had approved a draft bill which provides the legal framework to establish a system of reporting GHG emissions at the level of individual organisations (Pravitel'stvo Rossii, 2018). However, at the time of writing, the Ministry for Natural Resources was undertaking 'additional discussions' with the business community on the matter (TASS, 2018). It remains to be seen whether the bill will be submitted to parliament in its current form.

Stated measures for reducing GHG emissions are similar across the different companies surveyed, and usually fall within a company's broader energy saving plans. Gazprom, for example, prioritises GHG control as part of its corporate strategy, noting that 'energy saving practices and measures provide for the biggest GHG emission reduction at the gas transportation facilities' (Gazprom, 2016: 25). Rosneft claims to be 'working to lower its greenhouse gas emissions through initiatives provided in its Gas Investment Program and Energy Saving Program' (Rosneft, 2016: 74). Sakhalin Energy, in response to a stakeholder question about whether the company was planning further reductions in GHG emissions, claimed that 'the company is planning to maintain its current greenhouse gas emissions level at this stage. However, we are improving the energy efficiency of our assets as part of process optimisation. The implementation of these activities also has an effect on greenhouse gas emissions reduction' (Sakhalin Energy, 2016: 143).

A related issue concerns APG utilisation, or gas flaring, which is a major issue for the oil and gas industry. All eight companies identified APG utilisation as a priority, in line with government regulation, as noted above. A number of companies claim to have met the government's 95% utilisation target, including Tatneft (96.44%), Surgutneftegaz (99%) and Sakhalin Energy (96.3%).[6] Lukoil, which is 'approaching the target of 95%' (2015–6: 3), uses language to promote a 'green image' of the company, stating that, 'for over a decade, LUKOIL has been investing in APG utilisation projects, having started its APG flaring reduction effort long before legislators made it mandatory' (2015–6: 47). In terms of framing, however, little attempt is made to represent APG utilisation efforts as directly linked to climate change and emissions reduction efforts, and activities are reported within a wider framework of energy savings and efficient use of resources.

Energy saving and efficiency

All companies identified energy saving and efficiency as a key priority, and actively seek to promote their activities in this area as part of their environmental reporting. All companies have some kind of energy and resource-saving programme or policy. Rosneft's SR, for example, notes the company's success at meeting fuel and energy saving targets and states that 'lower operating costs resulting, in particular, from low fuel and energy consumption rates, along with higher production volumes are key for Rosneft to maintain its leading market positions' (Rosneft, 2016: 81). Similarly, Tatneft's AR points out that 'in the context of constantly growing tariffs of natural monopolies on energy resources, transportation services and increase of hydrocarbon production costs, the Company makes the comprehensive efforts to create maximum reserves for resource saving in all its business activities, including energy saving' (Tatneft, 2016: 52). The language used by companies to describe their efforts in this area demonstrates an economic framing of energy saving and efficiency.

A closely linked issue relates to industrial modernisation and an emphasis on new technology, which, similar to energy efficiency, corresponds to a market-based framing of climate change. While the Russian oil and gas companies surveyed did not draw direct links with climate policy, all AR and SR used language which focused on technology and modernisation. Igor Sechin's statement in Rosneft's SR highlights this point. Sechin, as CEO of the company, notes that 'striving to become a technology leader in the global energy sector, Rosneft consistently drives innovation by deploying new technologies, upgrading existing production processes and coming up with other initiatives' (Rosneft, 2016: 4). Lukoil, too, is 'focused on generating ideas, searching for and using new technologies (including those specifically aimed at reducing the consumption of natural resources), materials and energy with the highest output possible' (Lukoil, 2015–6: 21). Gazprom draws attention to its 'Innovative Development Program' which aims at the 'continuous improvement of technology' and 'technological leadership' to improve environmental safety and energy efficiency within the organisation (Gazprom, 2016: 64). Finally, Novatek and Rosneft both also make note of their involvement in best available technology 'technical working groups' (Novatek, 2016c: 26; Rosneft, 2016: 69), designed to develop guidelines for each industrial sector (Martus, 2017a: 279). The introduction of the 'best available technology' into Russian legislation has been a key development in recent years, and aims to use technology to improve environmental outcomes in industrial processes (Martus, 2017a).

There appears to be only minor interest amongst the companies sampled in renewable energy and investment. As noted above, government regulations are in place in Russia to encourage use of renewables (see Korppoo and Kokorin, 2017: 121), but with only limited effect so far. Lukoil, Novatek and Gazprom all make note of their use of renewables, with Lukoil reporting that 4% of 'total energy generated by the Company's power facilities is from renewables' (Lukoil, 2015–6: 20); and Novatek noting their use of solar panels and wind turbines along their pipelines (Novatek, 2016c: 77). Gazprom comments that it 'supports the use of alternative energy sources where it is economically and technically feasible', and provides a breakdown of generation by source (Gazprom, 2016: 49). However, none of the companies surveyed focus significant attention on the issue, suggesting that renewable energy use and investment is not a core part of corporate discourse on climate for oil and gas companies.

Discussion

A key issue to emerge from the analysis is how these industry narratives on climate change fit within a broader Russian and international context. That is, is the oil and gas industry exceptional in contrast with other sectors in

Russia? How does the industry's framing of climate change compare with the international industry discourse on climate change?

Within the SR, ER and AR documents analysed, corporate reporting on climate policy and climate-related policies was limited in comparison with other aspects of environmental action. When it comes to environmental protection, each of the companies surveyed provide extensive information about a range of issues, including waste and water management, and biodiversity conservation. They also all have some form of stand-alone policy on environmental protection, in contrast with climate. These documents or statements are used to set targets and identify key environmental tasks. As an example, Surgutneftegaz lists nine 'basic principles' in its environmental policy, including environmental monitoring, rational use of resources based on the implementation of 'innovative environmentally efficient and resource-saving technologies' (Surgutneftegaz, 2016: 6). The company has a list of responsibilities to implement these principles, including tasks such as adhering to federal legislation, assessment of environmental risks and impact assessments, and improving environmental safety (Surgutneftegaz, 2016: 7). Information is also provided about stakeholder engagement. Both Lukoil and Rosneft make note of joint activities with the WWF Russia for example. Lukoil, according to its report, signed a cooperation agreement with WWF, and conducted joint expeditions with the Marine Mammal Council to monitor walrus populations in the Barents Sea (Lukoil, 2015–6: 42). According to its SR, Rosneft consulted with the WWF on preserving biodiversity and reducing the impact of oil spills (Rosneft, 2016: 68). This does not mean these activities are necessarily anything more than 'environmental PR'; many of the tasks set by companies are vague with little clue given as to how they will be implemented, or they are tasks required by law anyway. AR and SR documents do not contain independent verification of company activities. However, what is demonstrated by this analysis is that the level of attention paid to the environment, in contrast with climate issues, is significantly higher.

While extensive research across economic sectors in Russia is lacking, a study by the present author of Russia's metals and mining sector provides a useful comparison (Martus, 2018). Similar to the oil and gas sector, the majority of metals and mining companies surveyed had a focus on energy saving and efficiency. As noted above, energy saving and efficiency measures have been a core focus of government climate policy efforts, and so we can consider oil and gas framing of the issue to be closely aligned with both the government and the metals and mining sector on this issue.

Beyond energy efficiency, however, differences emerge. The metals and mining companies were characterised by strong variation within the sector in terms of response to climate change and associated policy developments. While some companies were not supportive of action on climate, and either actively resisted or were found to be disengaged from the political discussion, a number of companies took a proactive role and sought to lobby the government to introduce GHG mitigation regulation and encourage greater Russian participation in international policy debates (Martus, 2018: 9). Aluminium company Rusal, and its oligarch owner Oleg Deripaska, was particularly active in this regard. Several companies had position statements broadly supportive of domestic and international efforts to address climate change, and joined domestic or international business partnerships such as the Climate Partnership of Russia or the UN Global Compact 'Caring for Climate', aimed at coordinating business commitments on climate (4–5). This level of activity was not found amongst the oil and gas companies examined. As noted above, Lukoil was the only company to have an official position on climate change and the only one to state that it supported international efforts to combat climate change. However, even this appeared limited, and beyond expressing support, Lukoil did not appear to play an active role in climate policy debates within Russia or internationally. On the whole, the variation present within the metals and mining sector was not found to be as significant in the case of Russian oil and gas companies, with the majority of them displaying very limited interest in climate policy and GHG mitigation efforts.

International comparisons are hampered by a relative lack of sector specific research. We would not expect a lack of interest or active disregard of climate mitigation policies to be unusual amongst global oil and gas majors. Although, as noted above, a number of works have pointed to the significant divergence in policy responses amongst oil and gas companies, particularly from the late 1990s. ExxonMobil, for example, is identified as a major opponent of climate policy, while other companies such as BP and later Shell took a more supportive stance (e.g. Skjærseth and Skodvin, 2001; van den Hove et al., 2002; Pulver, 2007). However, we would ultimately expect fossil fuel companies, both in Russia and globally, to be some of the most resistant to climate policy, given that climate change mitigation efforts therefore pose a threat to the operations of these companies.

Concluding remarks

This chapter has focused on industry and its understanding of climate change, as an important part of the broader discourse on climate change within Russia. As examined, all of the oil and gas majors present their plans to reduce GHG emissions and air pollutants, and outline measures to reduce gas flaring. Energy efficiency and resource-saving measures are explained in each report, and there is an emphasis on technology and industrial modernisation as a way of addressing the challenges presented by climate change and protecting the environment more broadly. The analysis provides clear evidence on an industry discourse on climate very much set within a market-based framework. It also highlights the lack of policy ambition on the part of the government, with delays to GHG reporting requirements and policies to encourage renewables that appear to be of limited interest for oil and gas companies.

So, while climate-related issues such as energy saving and efficiency do receive attention, there are clear limits on the extent to which oil and gas companies are prepared to engage. As we have seen, there is only very limited direct discussion of climate change, with only one company surveyed issuing a stand-alone position on climate. There was little to no engagement on domestic or international policy developments, in contrast to some companies within the Russian metals and mining sector as noted. While some attention was given to maintaining a 'green' reputation, the evidence suggested that climate was peripheral to this image. Finally, few companies gave consideration to the risk posed to their operations by a changing climate. For the most part, climate change seems to be of low importance for Russia's oil and gas sector.

Notes

1 For more detail on the sector, see Martus (2017b: 89–91).
2 For a comparison of the environmental policies of Rosneft and Gazprom, see Martus (2017b: 95–6).
3 $1US = 57.73 RUB as of December 2017
4 Japan represents an important export market for the company, accounting for 67% of LNG sales and 36% if oil blend sales in 2016 (Sakhalin Energy, 2016: 25).
5 In the sample selected here, all companies with the exception of Gazprom and Surgutneftegaz, used the Global Reporting Initiative (GRI) standards.
6 Tatneft (2016: 267); Surgutneftegaz (2016: 3); Sakhalin Energy (2016: 73).

References

Allen, M. and Craig, C. (2016) 'Rethinking corporate social responsibility in the age of climate change: a communication perspective', *International Journal of Corporate Social Responsibility*, vol 1, no 1, DOI 10.1186/s40991-016-0002-8.

Andonova, L.B. (2008) 'The climate regime and domestic politics: the case of Russia', *Cambridge Review of International Affairs*, vol 21, no 4, pp. 483–504.

Bashneft (2015) 'Otchet ob ustoichivom razvitii', www.bashneft.com/files/iblock/faa/20160617_bn_our2015_web_rus.pdf.

Bulkeley, H. and Newell, P. (2015) *Governing Climate Change*. Routledge, London.

Davydova, A. (2017) 'Parnikovye gazy vpisyvaiut v zakon: kompanii obiazhut otchityvat'sia ob ikh vybrosakh', *Kommersant*, 1 March, www.kommersant.ru/doc/3229089.

Ekspert.ru. (2017) 'Ekspert 400: reiting krupneishikh rossiiskikh kompanii', http://expert.ru/dossier/rating/expert-400/.

Gazprom. (2016) 'PJSC Gazprom Environmental Report 2016', www.gazprom.com/f/posts/44/307258/gazprom-ecology-report-2016-en.pdf.

Gazprom. (n.d.) 'O Gazprome', www.gazprom.ru/about/.

Gustafson, T. (2012) *Wheel of Fortune: The Battle for Oil and Power in Russia*, Belknap Press, Cambridge.

Hajer, M. (1995) *The Politics of Environmental Discourse: Ecological Modernization and the Policy Process*. Oxford University Press, Oxford.

Hoffman, A. (2005) 'Climate change strategy: the business logic behind voluntary greenhouse gas reductions', *California Management Review*, vol 47, no 3, pp. 21–46.

International Energy Agency (IEA). (2014) *Russia 2014*, IEA Publications, www.iea.org/publications/freepublications/publication/russia-2014-energy-policies-beyond-iea-countries.html.

Jones, C. and Levy, D. (2007) 'North American business strategies towards climate change', *European Management Journal*, vol 25, no 6, pp. 428–440.

Kolk, A. and Pinkse, J. (2007) 'Multinationals' political activities on climate change', *Business & Society*, vol 46, no 2, pp. 201–228.

Korppoo, A. (2007) 'Joint implementation in Russia and Ukraine: review of projects submitted to JISC', *Climate Strategies Briefing Paper*, October, http://climatestrategies.org/publication/joint-implementation-in-russia-and-ukraine-review-of-projects-submitted-to-jisc/.

Korppoo, A. and Kokorin, A. (2017) 'Russia's 2020 GHG emissions target: emission trends and implementation', *Climate Policy*, vol 17, no 2, pp. 113–130.

Levy, D. (2005) 'Business and the evolution of the climate regime: the dynamics of corporate strategies', in D. Levy and P. Newell (eds.), *The Business of Global Environmental Governance*. MIT Press, Cambridge, pp. 73–104.

Levy, D. and Kolk, A. (2002) 'Strategic responses to global climate change: conflicting pressures on multinationals in the oil industry', *Business and Politics*, vol 4, no 3, pp. 275–300.

Livesey, S. (2002) 'Global warming wars: rhetorical and discourse analytic approaches to ExxonMobil's corporate public discourse', *The Journal of Business Communication*, vol 39, no 1, pp. 117–148.

Lukoil. (2015–6) 'Lukoil Group Sustainability Report 2015–2016', www.lukoil.com/InvestorAndShareholderCenter/ReportsAndPresentations/SustainabilityReport.

Lukoil (n.d, -a) 'Health, safety and environment policy implemented by the open joint stock company "Oil Company Lukoil" in the twenty-first century', www.lukoil.com/Responsibility/SafetyAndEnvironment/HSEManagementSystem/HSEPolicy.

Lukoil (n.d, -b) 'Lukoil's stance on climate change', www.lukoil.com/Responsibility/SafetyAndEnvironment/Ecology/GasEmissionRegulation.

Lukoil (n.d, -c) 'Pozitsiia PAO "Lukoil" po Probleme Izmeneniia Klimata', www.lukoil.ru/Responsibility/SafetyAndEnvironment/Ecology/GasEmissionRegulation.

Mahoney, L.S., Thorne, L., Cecil, L. and LaGore, W. (2013) 'A research note on standalone corporate social responsibility reports: signaling or greenwashing?' *Critical Perspectives on Accounting*, vol 24, no 4–5, pp. 350–359.

Martus, E. (2017a) 'Contested policymaking in Russia: industry, environment, and the "best available technology" debate', *Post-Soviet Affairs*, vol 33, no 4, pp. 276–297.

Martus, E. (2017b) *Russian Environmental Politics: State, Industry and Policymaking*. Routledge, Abingdon.

Martus, E. (2018) 'Russian industry responses to climate change: the case of the metals and mining sector, *Climate Policy*, https://doi.org/10.1080/14693062.20 18.1448254.

Meckling, J. (2011) 'The globalization of carbon trading: transnational business coalitions in climate politics', *Global Environmental Politics*, vol 11, no 2, pp. 26–50.

Michelon, G., Pilonato, S. and Ricceri, F. (2015) 'CSR reporting practices and the quality of disclosure: an empirical analysis', *Critical Perspectives on Accounting*, vol 33, pp. 59–78.

Ministerstvo Finansov Rossiiskoi Federatsii. (2018) 'Ezhegodnaia informatsiia ob ispolnenii federal'nogo biudzheta' (dannye ot 1 ianvaria 2006 g.), www.minfin. ru/ru/statistics/fedbud/#.

Newell, P. and Paterson, M. (1998) 'A climate for business: global warming, the state and capital', *Review of International Political Economy*, vol 5, no 4, pp. 679–703.

Novatek. (2016a) 'Annual report', www.novatek.ru/en/investors/reviews/.

Novatek. (2016b) 'Politika PAO "Novatek" v oblasti okhrany okruzhaiushchei sredy promyshlennoi bezopasnosti i okhrany truda', approved 25 April.

Novatek. (2016c) 'Sustainability report', www.novatek.ru/en/development/.

Pravitel'stvo Rossii. (2018, 9 January) 'Komissiia po zakonoproektnoi deiatel'nosti odobrila zakonoproekt o sisteme otchetnosti ob ob"emakh vybrosov parnikovykh gazov', http://government.ru/dep_news/30901/.

Pulver, S. (2007) 'Making sense of corporate environmentalism', *Organization & Environment*, vol 20, no 1, pp. 44–83.

Rosneft. (2016) 'Rosneft sustainability report', www.rosneft.com/Development/ Sustainability_Reports/.

Sakhalin Energy. (2016) 'Sustainable development report 2016', www.sakhalinen ergy.ru/media/user/otchety/GRI_report_2016_eng.pdf.

Skjærseth, J.B. and Skodvin, T. (2001) 'Climate change and the oil industry: common problems, different strategies', *Global Environmental Politics*, vol 1, no 4, pp. 43–64.

Surgutneftegaz. (2016) 'OJSC Surgutneftegas environmental report', www.surgut neftegas.ru/en/ecology/reports/.

TASS. (2018, 16 February) 'Sergei Donskoi: govorit' o stabilizatsii rynka nefti rano', http://tass.ru/forumsochi2018/articles/4962147.

Tatneft. (2016) 'Tatneft Company annual report', www.tatneft.ru/for-shareholders/ information-disclosure/annual-report/?lang=en.

Taylor, C. (2013) 'The discourses of climate change', in T. Cadman (ed.), *Climate Change and Global Policy Regimes: Towards Institutional Legitimacy*. Palgrave Macmillan, Basingstoke, pp. 17–31.

UNFCCC. (2015) 'GHG emission profiles: Russian federation', http://unfccc.int/files/ghg_emissions_data/application/pdf/rus_ghg_profile.pdf.

van den Hove, S., Le Menestrel, M. and de Bettignies, H. (2002) 'The oil industry and climate change: strategies and ethical dilemmas', *Climate Policy*, vol 2, no 1, pp. 3–18.

Wright, C. and Nyberg, D. (2015) *Climate Change, Capitalism, and Corporations: Processes of Creative Self-Destruction.* Cambridge University Press, Cambridge.

Index